liberation biology

liberation biology

the scientific and moral case for the biotech revolution

ronald bailey

Prometheus Books
59 John Glenn Drive
Amherst, New York 14228-2197

Published 2005 by Prometheus Books

Inquiries should be addressed to
Prometheus Books
59 John Glenn Drive
Amherst, New York 14228–2197
VOICE: 716–691–0133, ext. 207
FAX: 716–564–2711
WWW.PROMETHEUSBOOKS.COM

09 08 07 06 05 5 4 3 2 1

Library of Congress Cataloging-in-Publication Data

Bailey, Ronald.
Liberation biology : the scientific and moral case for the biotech revolution / by Ronald Bailey.
p. cm.
Includes bibliographical references and index.
ISBN 1–59102–227–4 (hardcover : alk. paper)
1. Biotechnology—Social aspects. 2. Genetic engineering—Social aspects. 3. Bioethics. I. Title.

TP248.23.B35 2005
303.48'3—dc22

2005005875

Printed in the United States of America on acid-free paper

For my love, Pamela

"We could be free of an infinitude of maladies both of body and mind, and even also possibly the infirmities of age, if we had sufficient knowledge of their causes, and of all the remedies with which nature has provided us."

—René Descartes

CONTENTS

PREFACE

BRAVE NEW WORLD RECONSIDERED

If I could have given this book a proper nineteenth-century descriptive title, it would have been something like *Liberation Biology: The Scientific and Moral Case for the Biotech Revolution; Or, Why You Should Relax and Enjoy the Brave New World.*

The grim forecasts of the future in Aldous Huxley's *Brave New World* and George Orwell's *1984* exercised great influence over the cultural and political climate of the twentieth century. Both books sketched out gruesome visions of humanity enslaved by technologically advanced totalitarian despotisms.

In Orwell's book, Big Brother rules Oceania by means of pervasive surveillance combined with crude but effective brainwashing techniques. "Thoughtcrimes" are pitilessly corrected by commissars at the Ministry of Love. Oppression is justified as necessary to prosecute the perpetual war against either Eastasia or Eurasia. Citizens' access to information is strictly limited and the past can be changed on command. Soviet communism clearly informed Orwell's depiction of the future.

The collapse of Soviet communism and the end of the cold war at the conclusion of the twentieth century has lessened the force of Orwell's dark imaginings. The future doesn't belong to Big Brother and his minions, after all. However, as Orwell's telescreen tyranny has faded, Huxley's bleak fable of an omnipotent "benevolent" world state populated by eugenically generated castes looms ever larger in our cultural imagination and political debates.

"Just give us the technological imperative, liberal democratic society, compassionate humanitarianism, moral pluralism, and free markets, and we can take ourselves to a Brave New World all by ourselves—and without even deliberately deciding to go," declared Leon Kass, chairman of the President's Council on Bioethics in 2001. He further urgently warned, "In case you had not noticed, the train has already left the station and is gathering speed, but no one seems to be in charge."[1]

Liberal Boston University bioethicist George Annas has warned that it is time to develop a "species consciousness" to protect the human race against the dystopia described in Huxley's *Brave New World*.[2] In his first televised speech to the nation in which he placed limits on federal support for research on human embryonic stem cells, President George W. Bush warned, "We have arrived at that brave new world that seemed so distant in 1932, when Aldous Huxley wrote about human beings created in test tubes in what he called a 'hatchery.'"[3]

Clearly, contemplating some aspects of modern biotechnology has unhinged the imaginations of some of our prominent policy intellectuals and political leaders, linking current biomedical therapies in their minds with some of the ghastly technologies described in *Brave New World*. Kass in particular makes the connection clear when he warns: "Some transforming powers are already here. The Pill. In vitro fertilization. Bottled embryos. Surrogate wombs. Cloning. Genetic screening. Genetic manipulation. Organ harvesting. Mechanical spare parts. Chimeras. Brain implants. Ritalin for the young, Viagra for the old, Prozac for everyone. And, to leave this vale of tears, a little extra morphine accompanied by Muzak."[4]

But is biotechnological progress pushing humanity down the slippery slope to a Brave New World? Absolutely not.

Huxley's fictional Brave New World is filled with people who are merely interchangeable cogs in the great world state. They feel no great passion and experience no inspiration, living lives of happy, thoughtless mediocrity. Artistic and technological creativity are long dead. If modern biotechnology were in fact impelling humanity into this flat, soulless future, all people of goodwill should certainly oppose its further development. Fortunately, bioconservatives like Kass and Annas are wrong—the future toward which the biotech revolution is taking humanity is, in fact, almost the exact opposite of the Brave New World.

Huxley's eugenic civilization seven hundred years in the future, although "benevolent," is a centrally planned global state ruled by World Controllers. Its motto is: "Community, Identity, Stability." All citizens are produced in "hatcheries," where they grow from embryos to infants in bottles lined with porcine peritoneum and are fed with blood surrogate. The lower castes are created by means of a cloning process called "bokanovskification." Keep in mind that bokanovskification does not involve engineering genes—all embryos start with normal heredity. The caste into which a citizen is born depends entirely on environmental influences. The fetuses destined to become lower caste workers are denied oxygen, bombarded with x-rays, and so forth in order to stunt their mental and physical capacities and thus fit them for their subservient roles in society. The goal is to create "[s]tandard men and women; in uniform batches."

Besides being standardized as fetuses, the mental and emotional lives of the citizens of Huxley's world are shaped entirely by means of a thoroughly regimented system of conditioning. Each caste gets only the training for work and play that is appropriate to its status. Citizens have no choices about careers, hobbies, likes and dislikes—or anything else, really—but they don't care because in the best behaviorist tradition, they are trained to like whatever it is they get. And if there happens to be a momentary disappointment, there is always the bliss of soma to smooth things over. Huxley's Brave New World would be hell on earth.

But Huxley's fantasy has nothing to do with the modern biotechnology revolution.

Look again at the list of currently available technologies Kass apparently thinks are already dehumanizing us. What those modern technologies all have in common is that they give *individuals* more choices and options over how to enhance their lives, improve their health, and bear children. Kass complains about "harvesting organs," but isn't it morally laudatory to donate one's organs for use as transplants? He warns against "mechanical spare parts," but what could possibly be immoral about an artificial knee or hip? People are certainly not choosing to use biomedical therapies such as the birth control pill, in vitro fertilization and even Viagra so that they can produce "standard men and women in uniform batches." Unlike Huxley's fictional bokanovskification or conditioning regimens, none of the modern biomedical technologies cited by Kass diminish the life prospects of any individual. In fact, they do the opposite. Many enable people who would otherwise be "dehumanized" by disease, disability, or death to survive and flourish. Others allow people to bear children and manage their reproductive lives.

Look again too at the list of modern institutions that Kass fears are propelling a hapless humanity toward Huxley's dehumanized future. Surely he cannot be implying that in order to avoid that horrific fate that we must halt technological progress, overthrow liberal democratic society, dispense with compassion, compel a single moral code, and impose central planning on the economy? That would be a prescription for establishing Huxleyian "Community, Identity, Stability."

Liberation Biology will show that the bioconservative fears are vastly exaggerated, that their ethical objections to biotechnological progress are largely misconceived, and that rather than diminishing human dignity and liberty, the biotech revolution will instead enhance and enlarge them. In the twentieth century, liberation theology was a spiritual movement aimed at helping humanity to overcome political and economic oppression. In the twenty-first century, liberation biology is the earthly quest to overcome the physical and mental limitations imposed on us by nature, enabling us to flourish as never before. The true Brave New World toward which humanity's growing biotechnological prowess is aiming is not one populated by regimented clones, but one in which more and more individuals can exercise enhanced intellectual, creative, and physical capacities while being liberated from the immemorial curses of disease, disability, and early death.

ACKNOWLEDGMENTS

First, I must thank my colleagues at *Reason* magazine and the Reason Foundation for their support and backing of this project. My friend and editor, Nick Gillespie, has been especially encouraging and patient during the past year. Thanks also to David Nott, president of the Reason Foundation, for standing behind this project. I also want to thank my friend and colleague Brian Doherty for his valuable suggestions for improving this book.

I am most particularly grateful for generous and timely support from the Richard Lounsbery Foundation, the Bruce and Giovanna Ames Foundation, and the Alliance for Aging Research.

Thanks to my agent, Ted Weinstein, for finding a good home for this book. Speaking of good homes, I truly appreciate the patience of my editor Steven L. Mitchell at Prometheus Books.

Among the many others whom I must thank for championing me and my work are Virginia Postrel, Gregory Stock, Daniel Perry, Charles Harper, and Nick Schulz. Finally, and most important, I am forever indebted to the scores of researchers who shared their time, their scientific findings, their philosophical insights, and their hopes for the future of humanity with me. Without them and their work, our aspirations for liberating humanity from the ravages of disease, hunger, and old age would remain vain fantasies.

INTRODUCTION

BIOPOLITICS

Fight of the Century

By the end of the twenty-first century, the typical American may attend a family reunion in which five generations are playing together. And great-great-great grandma, at 150 years old, will be as vital, with muscle tone as firm and supple, skin as elastic and glowing, as the thirty-year-old great-great-grandson with whom she's playing touch football.

After the game, while enjoying a plate of greens filled with not only a solid day's worth of nutrients, but medicines she needs to repair damage to her aging cells, she'll be able to chat about some academic discipline she studied in the 1980s with as much acuity and depth of knowledge and memory as her fifty-year-old great-granddaughter who is studying it right now.

No one in her extended family will have ever caught a cold. They will have been from birth immune to such shocks as diabetes and Parkinson's disease. Their bodies will be as athletic as a human is capable of being, with brains as clever and capable of learning and retaining knowledge and training as human brains can be.

Her granddaughter, who recently suffered an unfortunate transport accident, will be sporting new versions of the arm and lung that got damaged in the wreck. She'll join in the touch football game, as skilled and energetic as anyone else there. Infectious diseases that terrifed us in the early twenty-first century, such as HIV/AIDS and severe acute respiratory syndrome (SARS), will be horrific historical curiosities for the family to chat about over their plates of superfat farm-raised salmon, as tasty and nutritious as any fish any human has ever eaten. Surrounding them, though few of them will think much of it with all their health, vitality, and riches, will be a world that's greener and cleaner, more abundant in natural vegetation, with less of an obvious human footprint, than the one we live in now. And it's not only this family that will enjoy all these benefits—nearly everyone they work with, socialize with, or ever meet will enjoy them as well. It will be a remarkably peaceful and pleasant world even beyond their health and wealth—antisocial tendencies, crippling depression, will all be managed—by individual choice—through new biotech pharmaceuticals and even genetic treatments.

This idyllic fantasy is more than realistic, given reasonably expected breakthroughs and extensions of our knowledge of human, plant, and animal biology, and mastery of the techniques—known as biotechnology—to manipulate and adjust those biologies to meet human needs and desires.

Although this vision would strike most people of goodwill and hope as encouraging, and devoutly to be wished for, an extraordinary political and intellectual coalition of left-wing and right-wing bioconservatives has come together to resist the biotechnological progress that could make that vision a reality for the whole human race. According to longtime antibiotechnology activist Jeremy Rifkin, "The biotech era will bring with it a very different constellation of political visions and social forces just as the industrial era did. The current debate over cloning human embryos and stem cell research is already loosening the old alliances and categories. It is just the beginning of the new biopolitics."[1] Unfortunately, he's right. For example, Rifkin has teamed with William Kristol, editor of the conservative *Weekly Standard*—a man who opposes him on almost every other public policy issue—to coauthor editorials calling for a ban on human cloning, one of the keystones of biotech—even cloning to produce stem cells that would be perfect transplants for repairing damaged hearts and spinal cords.[2] The abstract ideological objections of men such as Rifkin and Kristol promise prolonged misery for tens of millions of real, living people.

This transideological coalition of bioconservatives fears that the biotechnology revolution will usher in what they call "posthumanity" or "transhumanism." "The most significant threat posed by contemporary biotechnology is the possibility that it will alter human nature and thereby move us into a 'posthuman' stage of history," writes Francis Fukuyama in *Our Posthuman Future: Consequences of the Biotechnology Revolution.*[3] Fukuyama is a political science professor at Johns Hopkins University and, alas, a member of the President's Council on Bioethics. This growing bioconservative political movement aims to restrict scientific research, ban the development and commercialization of biotech products and procedures, and deny citizens access to the fruits of the biotech revolution—all because biotech innovation threatens their devoutly held notions of human nature, their social and political views, and their ideas of proper community control.

In February 2001, at the annual meeting of the American Association for the Advancement of Sciences, Francis Collins, head of the Human Genome Project, predicted, "Major anti-technology movements will be active in the U.S. and elsewhere by 2030." Collins's prediction is off by three decades. The closer the more dazzling promises of new biotechnologies come to reality, the more they incite opposition and, in some cases, horror. "We're embarking on a 100-year war about this issue," declared former US senator Larry Pressler in 2000 at the United Nations World Forum session titled "Advances in Biotechnology for the Human Genome and the Human Being: Technology, the Market and Responsible Citizenry." Another World Forum participant, philanthropist Walter Link described the current controversies over genetically enhanced crops as "a low-level form of civil war."[4]

The opposition to this rich, abundant human future is already on the march. On the Left, activists crowd the streets from Seattle to Johannesburg to protest the development of genetically enhanced crops. Rifkin asserts that biotech advances violate "the boundaries between the sacred and the profane" and demands "a strict global moratorium, no release of GMOs (genetically modified organisms) into the environment."[5] Rifkin is now ably assisted by his disciple, environmentalist writer Bill McKibben, who despairs, "I think genetically engineering our children will be the worst choice human beings ever make."[6] Marcy Darnovsky from the Center for Genetics and Society and Tom Athanasiou from EcoEquity claim, "The techno-eugenic vision urges us, in case we still harbor vague dreams of human equality and solidarity, to get over them." They fear that biotechnology will "allow inequality to be inscribed in the human genome."[7] Richard Hayes, director of the Center for Genetics and Society, declares that the movement opposing human genetic engineering "will need to be of the same intensity, scope, and scale as the great movements of the past century that struggled on behalf of working people, anticolonialism, civil rights, peace and justice, women's equality, and environmental protection."[8] And bioethicist Daniel Callahan, one of the cofounders of the Hastings Center, which is arguably the world's first bioethics think tank, contends that using biotechnology to extend human life spans beyond seventy years is inherently illegitimate.

But objections to extending the exercise of human liberty to our most intimate biological realities are rampant on the Right as well. The most prominent bioconservative is Leon Kass, from the University of Chicago, who heads up the President's Council on Bioethics. Kass fiercely opposed in vitro fertilization in the 1970s and now opposes any type of human cloning and all future genetic intervention in the human germline. Kass warns that biotechnologists who say they "merely want to improve our capacity to resist and prevent diseases, diminish our propensities for pain and suffering, decrease the likelihood of death" are deceiving themselves and us. Behind these modest goals, he says, lies a utopian project to achieve "nothing less than a painless, suffering-free, and, finally, immortal existence."[9] Human reproductive cloning must be banned based on what Kass calls the "wisdom of repugnance." Fukuyama approvingly notes, "In Europe, the environmental movement is more firmly opposed to biotechnology than is its counterpart in the United States and has managed to stop the proliferation of genetically modified foods there dead in its tracks. But genetically modified organisms are ultimately only an opening shot in a longer revolution and far less consequential than the human biotechnologies now coming on line."[10] And Adam Wolfson, editor of the neoconservative journal *Public Interest*, warns against the hubristic temptations offered by the biotech revolution: "So let's not fool ourselves: A sentiment less generous than education of the young drives the ambition to engineer smarter, cleverer beings. It is the desire for an even more complete mastery over nature."[11]

Fukuyama is one of the leading Right intellectuals in the war against what he calls transhumanism—described by Fukuyama as "a strange liberation movement" that wants "nothing less than to liberate the human race from its biological constraints."[12] But what he doesn't acknowledge with such scary-sounding pronouncements is that human history has always been all about liberating more and more people from their biological constraints. It's not as though most of us still live in our species' "natural state" as Pleistocene hunter-gatherers.

Human liberation from our biological constraints began when the first human sharpened a stick and used it to kill an animal for food. Further liberation from biological constraints followed with fire, the wheel, domesticating animals, agriculture, metallurgy, city building, textiles, information storage by means of writing, the internal combustion engine, electric power generation, antibiotics, vaccines, transplants, and contraception. In a sense, *the* goal toward which humanity has been striving for millennia has been to liberate ourselves, by extending our capacities, from more and more of our ancestors' biological constraints.

What is a "human capacity" anyway? Oxford University biologist Richard Dawkins has promulgated the notion of an extended phenotype. Genes not only mold the phenotypes, that is, the bodies of organisms, but also influence their behaviors. Some of those behaviors result in the shaping of inanimate objects, such as beaver dams and bird nests, that help organisms to survive and reproduce. Thanks to our knack for technological innovation, humanity has by far the largest extended phenotype of all creatures on earth. Our ancestors had no wings; now we fly. Our ancient forebears could not hear one another across one thousand miles; now we phone. And our Stone Age progenitors averaged twenty-five years of life; now we live seventy-five.

But will we be allowed to use our ingenuity to go even further? The bioconservative drive to slow and even halt the biotech revolution has already achieved some notable successes. In the United States, the House of Representatives has twice voted overwhelmingly to criminalize all research on human cloning, even research aimed at producing stem cells for transplants, though the bill so far remains stymied in the US Senate. If it were to pass, cloning research would be punished by ten years in jail and a $1 million fine. Even Americans who, in the future, might go abroad for embryonic stem cell treatments to cure diseases such as Parkinson's or diabetes would be imprisoned for illegally importing stem cells. In 2002 the President's Council on Bioethics voted 10–7 in favor of a moratorium on therapeutic cloning research. As for reproductive cloning, the council was unanimously in favor of a permanent ban.

The war against biotech has advanced even further in Europe. There, biotech opponents have essentially banned genetically enhanced crops even though every scientific body that has ever evaluated them has found them to be completely safe to eat. The United Nations has adopted a Universal Declaration on the Human Genome and Human Rights that, among other things, forbids "practices which are contrary to human dignity, such as reproductive cloning of human beings."[13] The Bush administration's effort to get the UN General Assembly to adopt a global treaty banning all human cloning failed by just one vote in 2003. However, the Bush administration did persuade the General Assembly to adopt a resolution urging member states to enact legislation "to prohibit all forms of human cloning in as much as they are incompatible with human dignity and the protection of human life."[14]

Bioconservatives fear that biotechnological advances threaten to undermine human dignity, equality, and community control. In the following chapters, I will show how biotechnological treatments and developments will instead enhance human dignity by enabling more and more people to live flourishing lives free of disease, disability, and the threat of early death. Rather than exacerbating human inequality, in the long run, safe genetic engineering will enable parents to give their children beneficial genes that other children get naturally. This is a recipe for eliminating genetic inequalities, not perpetuating them. And despite fears about loss of community control, biotechnological advances will certainly remain subject to laws intended to ensure their safety and efficacy, just as other medical treatments and products are today.

A biotechnological future determined by the choices and decisions of individuals who want to use technology to help themselves and their families live richer, fuller lives clearly frightens bioconservatives on both the Right and the Left. "Though well-equipped [through biotech], we know not who we are or where we are going," Kass fearfully writes.[15] If we know not who we are, advances in biotech are helping us to understand the answer to that most important of questions. As for where we are going—well, the fact that we don't know is why we go. Over the horizon of human discovery Kass sees a territory marked, like the maps of yore, "Here Be Monsters." To avoid these monsters, as fanciful as the sea serpents mariners of old feared, Kass and other bioconservatives want humanity to stay quietly at home with its old conceptions, technologies, traditions, and limited hopes.

If we are allowed to use biotech to help future generations become healthier, smarter, and perhaps even happier, have we "imposed" our wills on them, as bioconservatives warn? Will we have deprived them of the ability to flourish as full human beings? To answer yes to these questions is to adopt Jean-Jacques Rousseau's view of humanity as a race of happy savages, sadly degraded by civilization. Previous generations have of course "imposed" all sorts of technologies and institutions on us. Thank goodness they did, because by any reasonable measure we are far freer, richer, better off than our ancestors. Our range of choices in work, spouses, communities, medical treatments, transportation—the list is endless—is incomparably vast. Like earlier technologies, biotech will liberate future generations from today's limitations and offer them a wider scope of freedom, more and better tools to live whatever life they desire.

This is the gift we can give them—if the decision makers in the worlds of policy and politics, whipped into a frenzy by bioconservative fears, don't try to prevent us from doing so.

Scientific facts in and of themselves will not resolve the controversies over biotech and what it can and should do. We must each examine our consciences and our experiences to find the answers. People who see human genes as the defining essence of humanity will object to stem cell research and a good deal else in the coming biotech revolution. On the other hand, people who see human beings as defined essentially by their minds will have fewer moral objections.

The following chapters will highlight the battles that are already turning the biotechnology century we have just entered into an ideological war zone. I will show how most bioconservative arguments for slowing and stopping biotechnological progress are shortsighted—and even immoral. After all, delaying technologies can kill people. If a cure for cancer that would otherwise have been available in 2010 is delayed until 2020 because of bioconservative opposition, tens of millions of people who would have been saved will be dead before the cure is found. I will make the general case that biotech research and the powerful new technologies that it engenders should be encouraged and embraced, not feared and resisted. A new pro-biotech politics must grow to counter the bioconservative ideology of stagnation. Ultimately I will argue that the decision to use or not use new biotechnological treatments involving stem cells, gene therapy, genetic diagnostics, preimplantation embryo testing, or genetic enhancements should rest with each citizen and family, not with a fearful claque of fundamentalists out to forbid their fellow citizens access to the future benefits of the biotechnology revolution.

In his famous essay "The End of History," Fukuyama declares that our generation is witnessing "the end point of mankind's ideological evolution and the universalization of Western liberal democracy as the final form of human government." Fair enough. But for Fukuyama, the end of history is a "sad time" because "daring, courage, imagination, and idealism will be replaced by economic calculation." He also laments that "in the post-historical period there will be neither art nor philosophy, just the perpetual caretaking of the museum of human history."[16] How ironic that Fukuyama now spends his time demonizing biotechnological progress and transhumanism, a nascent philosophical and political movement that epitomizes the most daring, courageous, imaginative, and idealistic aspirations of humanity.

In a sense, the battle over the future of biotech—and the future of humanity—is between those who fear what humans, having eaten of the Tree of Knowledge of Good and Evil, might do with biotech, and those who think that it is high time that we at long last also eat of the Tree of Life.

CHAPTER 1

FOREVER YOUNG

The Biology and Politics of Immortality

Goat testicle transplants. Elixirs of jade. Inhaling the breath of virgins. Injecting crushed dog gonads. Drinking radioactive waters.

These are just a few of the methods people have used in an attempt to lengthen their lives and renew their vitality. The oldest narrative known to humanity—the nearly five thousand-year-old Gilgamesh epic from ancient Sumeria—describes a quest for immortality and perpetual youth. Enkidu, bosom buddy of the semidivine King Gilgamesh, is killed for mocking the gods. The heartbroken king seeks the advice of Utnapishtim and his wife, the only two mortals to whom the gods have granted eternal life. Utnapishtim directs Gilgamesh to a certain waterweed that will restore his youth. Gilgamesh finds it but falls asleep, and a snake eats the weed. In the end, Gilgamesh realizes that the only immortality human beings can aspire to is making names for themselves as builders of cities.

This is, to say the least, unsatisfactory. As Woody Allen once put it, "I don't want to achieve immortality through my work. I want to achieve it through not dying."[1] Modern biomedical researchers, in the quest for the equivalent of Gilgamesh's waterweed, have made great progress in unraveling the mystery of aging. Physical immortality may not be in the immediate offing, but the day may come when death is radically postponed, if not fully optional.

However, the barriers to this goal are not just biological but political. Believe it or not, some of our most influential contemporary intellectuals are opposed to the idea of long, healthy lives. "The finitude of human life is a blessing for every individual, whether he knows it or not," declares Kass.[2]

From the political Left, Kass is joined in his opposition to abundant life by Daniel Callahan. "There is no known social good coming from the conquest of death," Callahan declared at a March 2000 conference on aging and life extension. "The worst possible way to resolve this issue is to leave it up to individual choice."[3]

Despite the ideological opposition, there's good reason for optimism on the scientific front for dramatically extended life spans. "The prospects of dramatically increasing human longevity are excellent," declares Steven Austad, a biologist at the University of Idaho. "Don't expect them tomorrow, but there will be major advances within the next fifty years." Austad expects twenty- to forty-year jumps in longevity to occur later in this century.[4]

"We see ourselves on the cusp of the second longevity revolution," agrees Jay Olshansky, a demographer at the University of Chicago. "Scientists are on the verge of discovering major secrets of aging."[5] Even the cautious President's Council on Bioethics acknowledges in its 2003 report *Beyond Therapy: Biotechnology and the Pursuit of Happiness* that "[i]t seems increasingly likely that something like age-retardation is in fact possible."[6]

The first longevity revolution occurred in the early twentieth century as infant mortality declined and infectious diseases were largely conquered; as a result, more young people now enjoy the opportunity to become old. The next longevity revolution, by contrast, will actually postpone old age.

Of the two longevity researchers, Olshansky is the pessimist. He bet Austad $500 million that there will no 150-year-old person alive and in fairly good shape in 2150. The bet will be financed through a trust fund endowed by $150 from each scientist. Given the math behind compound interest, the winner's heirs will receive a $500 million payout on January 1, 2150.

LOVE AND DEATH

Woody Allen might appreciate the science of longevity as well as its promised results. Sex and death, it turns out, are inextricably intertwined. "It doesn't pay to have a body that will last forever," notes Austad. "Evolution only cares about reproduction."[7] Every body harbors a line of immortal cells: the germ cells that produce eggs and sperm. In a sense germ cells use eggs and sperm to migrate from body to body down the generations, disposing of their worn-out carriers as they move on. Once your ovaries or testes are done with you, your gonads couldn't care less whether you live.

Here, then, is the definitive answer to the eternal question: Which came first, the chicken or the egg? The egg did. A chicken is merely an egg's way of making more eggs.

"When there's no future reproduction, then there's no reason for your survival," explains University of California, Irvine biologist Michael Rose.[8] Individuals are selected by nature so that they keep their health and their ability to evade predators and obtain sustenance long enough to get the next generation up to reproductive snuff. Once that job is done, nature throws our bodies into the dump to be recycled by worms.

In the case of human beings, evolution has selected for a set of genes that keep our bodies in pretty good shape long enough to mature sexually, produce progeny, and raise those progeny to sexual maturity. Time elapsed: about forty years. If a body invests a lot of energy in repairing itself, it will reduce the amount of energy it can devote to reproduction. That may be good for individual bodies, but, as previously noted, your germ cells have no interest in keeping you forever young.

So significant aging is a relatively new phenomenon. Throughout evolutionary history, once creatures began to falter in any way, they were eaten or dropped dead of disease—eaten by bacteria, as it were—so they never had a chance to get old. "As soon as an animal in the wild starts to slow down even a little bit, it's eliminated from the population very quickly," says Simon Melov, a researcher at the Buck Institute for Aging Research, a nonprofit organization devoted to aging research located in Novato, California. "Aging is not a natural state for us either," he adds. "The evidence we have indicates that 10,000 to 20,000 years ago, most people didn't live much past 30."[9] Today, the only creatures that actually experience aging are human beings and the animals they protect, such as pets and livestock. Leonard Hayflick, the San Francisco biologist who first discovered in the 1960s that human cells divide a limited number of times and then die, agrees: "Aging is a human artifact that in the natural world didn't occur. . . . Civilization has revealed a process that teleologically we were never designed to experience, that is, aging."[10]

Rose firmly established the evolutionary connection between sex and death by breeding fruit flies.[11] He selected only those flies that reproduced late in life and bred them with one another. The longer it took the insects to reproduce, the longer they lived. Rose now has flies that live 130 days instead of the usual forty. More significantly for useful longevity research purposes, the long-lived flies remain healthy and active up to the very end of their lives.

The sex-death connection was further bolstered by Cynthia Kenyon, a biologist at UC San Francisco and one of the world's leading researchers on aging. In 2002 Kenyon and her colleagues reported that she could double the life spans of nematode worms by removing their germline stem cells—the cells that produce eggs and sperm.[12] Kenyon thinks the germline stem cells affect life span by influencing the production of, or the response to, a steroid hormone that promotes longevity.

But what is aging, anyway? Cambridge University theoretical biogerontologist Aubrey de Grey defines aging as "a collection of cumulative changes to the molecular and cellular structure of the adult organism, which result from essential metabolic processes, but which also, once they progress far enough, increasingly disrupt metabolism, resulting in pathology and death."[13]

"The reason we age is because the processes that prevent things from breaking down stop working," explains Kenyon. "In principle, if you understood the mechanisms of keeping things repaired, you could keep things going indefinitely."[14] Aging results when certain thresholds of damage occur in the normal course of living. De Grey points out that without these cumulative changes in their bodies, forty-year-olds would have the same physical composition as twenty-year-olds. They would also have the same future life expectancy. De Grey continues, "The fact that forty-year-olds actually live, on average, nearly twenty years less than twenty-year-olds is due 100 percent to the fact that their bodies have accumulated subtle (and, thereafter, increasingly unsubtle) changes."[15]

Of course, many things besides aging kill people—primarily diseases and accidents. Then again, as we age we become increasingly vulnerable to those deadly shocks and diseases. In modern societies, disease plays a relatively small role in killing off younger people: If all the things that kill people in the United States before age fifty were eliminated, average life expectancy would increase by only three and a half years. Meanwhile, Olshansky has estimated that if all deaths from heart disease, cancer, and stroke were eliminated entirely, the average life expectancy in the United States would increase to between ninety and ninety-five years. In the nineteenth century the insurance actuary Benjamin Gompertz pointed out that after people reach their sexual peak, their chance of dying doubles every eight years or so. Thus, a thirty-five year-old is twice as likely to die at age forty-three and four times as likely to die at age fifty-one.

But there is no reason to think that life expectancy increases can't continue. Researchers reported in the April 29, 2002, issue of *Science* that life expectancy has been increasing at about two and a half years per decade for the past 160 years. Demographers such as Olshansky, they note, have been consistently wrong in predicting an upper limit to this trend. In 1928, for example, demographer Louis Dublin predicted that average life expectancy in the United States would never exceed 64.75 years. Today it is 77.6 years.

At this rate of improvement, the authors of the *Science* report conclude that "record [average] life expectancy will reach about 100 in six decades."[16] Still, as far as we know, the maximum human life span is the 122 years achieved by the cigarette-smoking Frenchwoman Jeanne Calment, who died in 1997. King's College London gerontologist Robert Weale has done some theoretical calculations, assuming the progressive decline of various biorepair mechanisms, and concludes that their repair efficiencies fall to zero by the twelfth decade of life. Weale notes that "if valid, the hypothesis indicates that an uphill struggle is likely to be needed to achieve [appreciable increases in human longevity]."[17]

Scientists are now in wide agreement that one of the primary causes, if not *the* cause, of aging is the damage done to our cells by free radicals. Also called reactive oxygen species, free radicals are atoms or molecules that have at least one unpaired electron, making them very chemically reactive. Lots of free radicals are created in cells through the normal processes of metabolism as they produce and utilize energy. As a side effect, these free radicals disrupt a cell's DNA and protein synthesis and repair mechanisms. The University of California, Berkeley biologist Bruce Ames has calculated that free radical versions of the oxygen atom damage the DNA inside every one of our cells some ten thousand times per day.[18]

Each time a cell replicates, it copies all of the billions of DNA base pairs that comprise its genome. Cells are very crowded and chemically energetic places, so miscopying sometimes occurs. Fortunately, evolution has devised molecular machines that can rapidly read and then correct most of the copying mistakes, keeping the cells to a fantastically accurate rate of one error per 50 million nucleotide replications. However, each time the repair mechanisms miss a mistake, it becomes encoded in the DNA—and the next time duplication occurs, the miscopied DNA is treated as correct. As a result, errors accumulate over time. Miscopied genes lead to the production of distorted proteins, which are inefficient when they work at all. The accumulated molecular damage causes a 0.5 percent decline per year in overall physical capacity after age thirty.

Antiaging researchers are focusing their attention increasingly on the damage free radicals cause in the tiny energy-producing organelles called mitochondria. Each cell contains thousands of these tiny power plants.

Like any other power plant, mitochondria produce not only energy but also wastes and pollution, including copious free radicals. As good as mitochondria are at mopping these up, some of them nevertheless get loose and damage the thirteen genes that code for proteins that constitute the tiny genomes at the heart of each mitochondrion. Free radicals create a cellular death spiral by mutating mitochondrial DNA, which in turn degrades their energy production and increases the production of more free radicals, refueling the cycle. Mitochondrial mutations and deletions are known to accumulate in a variety of tissues during aging in humans, rodents, and monkeys.

In 2004 the hypothesis that mitochondria play a big role in aging got an enormous boost through an elegant experiment carried out by a team of Scandinavian scientists. The scientists created mice with a defective proofreading and repair mechanism for their mitochondrial DNA. These "mutator mice" produced three to five times more mutations in their mitochondrial DNA than normal mice. What happened? They died early. Normal mice typically live two to three years, but these mice first showed signs of aging after only twenty-five weeks, and most lived less than a year. None survived more than sixty-one weeks. Furthermore, they prematurely showed many of the signs we associate with aging in humans and other mammals—weight loss, muscle wasting, hair loss, curvature of the spine, osteoporosis, anemia, reduced fertility, and heart enlargement.

Other processes going on in our bodies in the act of living also contribute to our eventual demise. In a sense, we are literally cooking ourselves to death. In cooking, sugars and proteins stick together to form tasty brown crusts like those on French toast. Our own metabolism is a form of low-temperature cooking that causes sticky sugars such as glucose to cross-link with proteins to create advanced glycation end products (AGEs). AGEs are biological junk that accumulate in cells, interfering with their function over time. For example, AGEs reduce the elasticity and flexibility of the collagen in our ligaments. They are also linked to diabetes and cardiovascular disease.

In a related process, our cells' recycling centers, called lysosomes, become clogged with cross-linked proteins and other cellular rubbish.

This cellular gunk is called lipofuscin. Lipofuscin-filled lysosomes slowly crowd and hinder other cellular functions. Inflammation is another hazard caused by free radicals. Inflammation occurs when our immune system cells drench invaders such as bacteria and viruses with free radicals to rip them apart. Sometimes the inflammatory attacks don't ratchet down when the threat is gone, and our immune cells keep pumping out free radicals. Chronic inflammation has been linked to many diseases associated with aging, including arthritis, arteriosclerosis, diabetes, Alzheimer's, and cancer.

Researchers Caleb Finch and Sarah Crimmins from the Andrus Gerontology Center at the University of Southern California recently made a fascinating finding: Exposure to infections in childhood boosts chronic inflammation levels in adults.[19] When the researchers checked people who suffered a higher-than-ordinary number of chronic infections in childhood, they found higher levels of various infection-fighting molecules such as C-reactive protein, interleukin-6, tumor necrosis factor, and so forth. It turns out that what is beneficial in the short term—a strong inflammatory response to defeat disease organisms in childhood—has negative effects over the long term, specifically increasing the chances of heart disease, diabetes, stroke, and cancer in late adulthood. Of course, in keeping with natural selection's theme of linking sex and death, the child saved from infection now has a chance to reproduce before succumbing to heart disease later in life.

As modern medicine and sanitation took hold, infants and children suffered many fewer infections. As a result, people living in modern societies experience much lower rates of heart disease and cancer than did people at the same ages in earlier generations. Finch and Crimmins report that "declining infection during the 20th century has been estimated to explain 11 to 24 percent of the reduction in late-life morbidity and mortality."[20] Scientists increasingly accept the idea that controlling inflammation has an important role in lengthening life by preventing heart disease, diabetes, and cancer. They often now advise taking nonsteroidal anti-inflammatory drugs such as aspirin. Cholesterol-lowering statins such as atorvastatin (Lipitor) also help control inflammation.

IMMORTALITY WHEN?

So what are your chances of living forever? First, the bad news. As Hayflick says, "There is no intervention that has been proven to slow, stop, or reverse aging. Period."[21] A position statement on human aging issued in 2002 by fifty-one of the world's leading researchers in the field of aging declared that "claim[s] that it is now possible to slow, stop or reverse aging through existing medical and scientific interventions . . . are as false today as they were in the past."[22]

Now the good news: Despite this, researchers are making a lot of progress. Even the fifty-one skeptical researchers agreed: "Most biogerontologists believe that our rapidly expanding scientific knowledge holds the promise that means may eventually be discovered to slow the rate of aging. If successful, these interventions are likely to postpone age-related diseases and disorders and extend the period of healthy life."[23]

The most promising immediate thing you can do to increase your chances of seeing your great-grandchildren is to stop eating so much. Calorie restriction (CR) is the only known technique for increasing the life spans of many different organisms. In 1935 Clive McCay, a professor of nutrition at Cornell, noted that if laboratory rats are fed only about two-thirds of the food they would freely choose to eat, their life spans increase by 40 to 50 percent. This result has been confirmed many times since then. Researchers think that semistarvation may make metabolic processes more efficient, producing fewer free radicals, and also perhaps boosting cells' DNA repair systems. Such boosted DNA repair appears to be an evolutionary adaptation that delays aging in undernourished animals in order to give them time to reproduce when food becomes more abundant.

The foremost advocate of this approach is probably the late UCLA biologist Roy Walford, whose Web site offers menus for those who want to pursue longer life by dieting themselves nearly to death. Walford died at age seventy-nine. People choosing to cut their calories to only twelve hundred per day may well increase their life spans, but CR practitioners' bodies look emaciated and their body temperatures are two or three degrees lower than normal, making them more sensitive to cold.[24] They generally have a reduced sex drive as well. As one wag put it: Calorie restriction may not make your life longer, but it will certainly make it feel longer.

Barbara Hansen, a diabetes researcher at the University of Maryland, has spent twenty years investigating the effects of calorie restriction on rhesus monkeys. The experiment has not run long enough to determine if the ultimate life spans of the calorie-restricted monkeys will be increased. But on several related issues, her findings are unequivocal: When she compares old calorie-restricted monkeys to old monkeys that eat what they want, the calorie-restricted ones have less heart disease, diabetes, and hypertension, and their cholesterol is lower. They're healthier.

There is no question that maintaining a reasonable weight prevents the onset of many illnesses associated with aging. But what if you find the notion of living without foie gras and pepperoni pizza unbearable? Is there hope for you? There may be. Hansen hopes that by starving monkeys she can pave the way to a pill providing the health benefits of calorie restriction while still allowing you to inhale all the ice cream and beer you want. She has identified a compound that affects the PPAR-delta receptor, which improves the body's response to insulin and glucose, mimicking the benefits of calorie restriction. The experimental drug, called GW501516 and developed by the biotech startup Ligand Pharmaceuticals, is now in early testing by GlaxoSmithKline. Researchers at the Salk Institute of Biological Studies have fed it to normal mice on a high-fat diet. The mice eat as much as they want and do not gain weight.[25] There are no data yet on how it might affect their life spans.

In the early 1990s Kenyon made the profound discovery that a single mutation in the daf-2 gene doubled worms' life spans. The daf-2 gene encodes a hormone receptor similar to the human receptors for the hormones insulin and insulin-like growth hormone-1 (IGF-1). The mutation causes the worms' cells to not respond to IGF-l. Later, when Kenyon removed the daf-2 mutants' germline cells, they lived for three months—six times longer than normal nematode worms. This is the equivalent to a human being living five hundred years.

The amazing thing about daf-2 mutant worms is that not only do they live a lot longer, they stay healthy a lot longer, too. Under the microscope, normal nematode worms near the end of their lives (fourteen days or so) hardly move and look positively ratty. In contrast, the daf-2 mutants remain active and sleek until the end of their much longer lives. Daf-2 mutants are also as fertile as normal worms. The reduced signaling of the daf-2 gene caused by the mutation causes a cascade of effects on about two hundred other genes that, in combination, increase longevity. It turns out that many of the downstream genes thus affected produce antioxidant proteins, antimicrobial proteins, and chaperone proteins that help repair or degrade other damaged proteins. However, there is one down side—daf-2 mutants thrive under a narrower range of temperatures than do normal worms. Outside the lab, in more unpredictable environments, normal short-lived worms out-compete daf-2 mutants by producing more offspring—sex and death again.

So scientists are able to boost the life spans of a bunch of tiny worms. Who cares? It turns out that evolution has highly conserved specific hormonal pathways—that is, very similar pathways have been maintained over time after the ancestors of worms and vertebrates split hundreds of millions of years ago. Thus, mammals, including human beings, have pathways similar to the IGF-1 pathway found in worms. "Maybe one day we will be able to tweak the insulin/IGF-1 systems in humans to produce many of the same benefits that we see in the worm," says Kenyon.[26]

MIT biologist Leonard Guarente recently showed that nematode worms with more than one copy of a gene called SIR2 live 50 percent longer than normal worms. This may explain why calorie restriction increases life span; SIR2 slows down the activity of various genes when a cell is being starved. The mammalian version of SIR2, called Sirt1, seems to increase longevity by lowering the levels of certain growth hormones or by preventing cells from dying through apoptosis.[27] (Apoptosis is programmed cell death, the body's normal method of disposing of damaged, unwanted, or unneeded cells.)

In June 2004 Guarante published a study in *Nature* showing that Sirt1 causes mice to shed fat. When the Sirt1 gene senses a sustained lack of food, it produces a protein that turns off the PPAR-gamma receptors that normally tell fat cells to keep storing fat. PPAR-gamma genes, like the

PPAR-delta genes studied by Hansen, are involved with fat and glucose metabolism. "The ability of fat cells to sense famine and release the fat is regulated by this gene," says Guarente. "We like to think this applies to people as well as mice, but we don't know for sure. If we could make this happen in people, it wouldn't just make them live longer; it might also help prevent diseases of aging, like cancer, diabetes and heart disease."[28]

Besides deactivating PPAR-gamma, David Sinclair at Harvard Medical School has found that Sirt1 also promotes cell survival by preventing the p53 tumor suppressor gene from ordering cells to commit suicide, and also by possibly boosting the DNA repair mechanisms inside cells.[29] "These interactions keep cells alive longer and buy more time for their DNA damage to be repaired," explains Sinclair.[30]

Bringing all the players together, it turns out that the IGF-1 and insulin also affect the Sirt1 pathway. When food is plentiful the body produces a lot of IGF-1 and insulin, which block Sirt1's ability to turn on the various cellular pathways that prevent cell suicide and promote cell longevity.

So Guarente, Kenyon, and others have identified several genes and molecular pathways that affect the life spans of various species. They believe that these same mechanisms will be found in people, providing targets at which to aim antiaging drugs. To facilitate the search for antiaging compounds, Guarente and Kenyon have joined forces to create a company called Elixir Pharmaceuticals.

In pursuit of a way to easily harness the calorie restriction/long life connection, the scientific hunt is on for experimental drugs called sirtuin activators (STACs) that will fool cells into thinking they're starving. The aim is to trigger the Sirt1 pathway and thus provide people with the health and longevity benefits of calorie restriction without having to actually keep to a superstrict diet—allowing us to have our cake and eat it, too. Scientists have found that resveratrol, a polyphenol antioxidant commonly found in red wine, is a powerful promoter of Sirt1 activity. When added to yeast cells, resveratrol lengthened their life spans by 70 percent and considerably extended the lives of human cells with damaged DNA in test tubes.[31]

Resveratrol may be partially responsible for the so-called French Paradox. The French diet is typically high in fat content. Other things being equal, that should significantly heighten the risk of cardiovascular disease. Yet the French suffer considerably less heart disease than do the British or Americans. Researchers speculate that the French habit of drinking red wine may be protecting them from the ill effects of their high fat diets. Unfortunately, resveratrol breaks down easily when exposed to oxygen, so researchers such as Guarente are searching for other, more stable molecules that can be turned in to effective drugs. Elixir Pharmaceuticals is already getting some competition since Sinclair has founded Sirtris Pharmaceuticals, which aims to find resveratrol mimics.

Calorie restriction research could also help biotechnologists discover the elusive biomarkers of aging—that is, the genetic and molecular changes in cells and tissues that cause or accompany aging. With the sequencing of the human genome and the advent of chip gene technologies, researchers can monitor the simultaneous actions of thousands of genes in tissues. The aim is to characterize the differences between gene activity in tissues from old people and that of young people. Already, chip tests have found that the gene expression of tissues taken from old calorie-restricted mice is similar to that found in young mice. Among other benefits, finding such biomarkers would be a major advance in testing the effectiveness of antiaging interventions. Right now, the only way to see whether a proposed longevity treatment works is to wait for people to die. That's obviously an unsatisfactory state of affairs.

CANCER AND TELOMERES VERSUS AGING

Aging, as it turns out, may simply be an unfortunate side effect of the body's otherwise laudable efforts to keep us alive long enough to reproduce. Lawrence Donehower at Baylor Medical College has recently found that genes helpful in guaranteeing a robust youth can be harmful in the long run.

Donehower genetically engineered mice so that they expressed a super powerful version of tumor-suppressing proteins produced by the p53 genes. Donehower hypothesizes that the protection offered by p53 comes at the cost of stimulating our immune systems to destroy over time the reserve of rapidly dividing stem cells that constantly replenish our tissues. As our stem cells are killed off, our tissues deteriorate. This is a case in which the effects of a single gene, p53, are pleiotropic. *Pleiotropic* means that a gene is responsible for a number of distinct and seemingly unrelated physical effects in the body. In this case, the dual effects are called antagonistic pleiotropy because they pull in opposite directions—p53 keeps us alive by protecting against cancer, but at the same time it destroys stem cells, which results in aging and eventual death. Although cancer free, Donehower's mice aged rapidly, with life spans 30 percent shorter than those of normal mice.

Hayflick discovered another kind of antagonistic pleiotropy in 1961. Hayflick found that human cells in vitro divide only fifty to eighty times, and then stop. For a while, some researchers thought this "Hayflick limit" might be the key to aging. It appears to be an evolved mechanism to prevent cancer. Cancer is the uncontrolled proliferation of cells, and as cells divide they often accumulate errors that predispose them toward becoming cancerous. If there is a limit to the number of times a cell can divide, that prevents cells from eventually mutating into cancer cells. So the body has essentially two ways to deal with damaged cells: It can tell them to stop dividing (senescence) or it can tell them to die (apoptosis). Both mechanisms evolved to keep cancer at bay until after an individual has reproduced.

Intrigued by Donehower's research, biologist Heidi Scrable at the University of Virginia created a mouse that overexpresses a short version of the p53 protein, called p44. P44 causes the tumor-suppressing p53 to become hyperactive. Scrable's p44 mice are small, about one-half the size of their normal littermates. They begin to look old at four months. Most of them are dead within a year. (Mice with normal p53 activity live two to three years.) Although they age and die prematurely, p44 mice are extraordinarily resistant to cancer.

How does hyper p53 promote premature aging? This is where we come back to Kenyon's research. Apparently, among other things, hyper p53 stimulates the production of IGF-1. Initially this seems like an odd result, since the p44 mice are small and IGF-1 is a growth hormone. In addition, IGF-1 is also known to stimulate the progression of cancer, yet the mice are cancer free. So what gives?

Judith Campisi, a cancer and aging researcher at University of California, Berkeley and the Buck Institute for Aging Research, believes that the hyper-p53 activity forces cells to stop dividing, that is, to become prematurely senescent. As noted earlier, cell senescence is one of the two ways organisms protect themselves against cancer; the other is to kill them off in an orderly manner called apoptosis. This is why p44 mice don't have cancer, yet they age rapidly. In addition, the p44 mice are about half the size of normal mice because the process of senescence is so vigorous that their tissues don't have a chance to grow to full size.

Senescence has another cause as well: telomeres, caps composed of repetitive DNA sequences at the ends of chromosomes. These function somewhat like the aglets on the ends of shoelaces, keeping the chromosomes from unraveling or attaching themselves to other chromosomes. Through a peculiarity in DNA replication, each time a cell divides, its daughter cells lose a tiny bit of their telomeres. Once the telomeres have eroded away, the cell stops dividing and becomes senescent.

However, two types of cells escape the various signals that tell them to stop dividing. One is the germ cells that are the progenitors of sperm and eggs. Their telomeres are restored by an enzyme called telomerase, making them essentially immortal. And cancer cells, too, use telomerase to restore their telomeres. Cancer cells can divide without limit, which is what makes them so deadly. The Geron Corporation is conducting research into how to use telomerase to fight cancer. The idea is that if you can turn off telomerase in the cancer cells, they will stop dividing and commit cellular suicide.

Some researchers have suggested that you could make normal cells immortal by getting them to produce telomerase, thus preventing the shortening of their protective telomeres. It seems logical that if telomere shortening causes cells to become senescent, lengthening them should rejuvenate them and the tissues in which they reside. Although not proven yet, it seems that "all aging is cell aging" according to antiaging researcher Michael Fossel, a clinical professor of medicine at Michigan State University.

Fossel can now make senescent cells young again by telomerizing them ex vivo, that is, by restoring the tips of chromosomes, which get shorter each time a cell divides. Cells with restored telomeres are rejuvenated and have normal gene expression. The goal is to figure out how to induce telomerase in our bodies. "If we could do that, nothing would have greater impact on human health and longevity," says Fossel. Fossel may be a bit overenthusiastic because, according to Barbara Hansen, there is little evidence that telomere shortening causes senescence on the organism level.

VITAMINS AND HORMONES

Since the damage caused by free radicals is so central to our decay, we have a far lower-tech way to potentially attack aging available to us: supplementing the body's own antioxidant compounds. The most popular antiaging regimen, practiced by millions, is to pop antioxidant vitamin and mineral pills. Unfortunately, there is no firm scientific evidence that gobbling down megadoses of such supplements actually increases life spans or even much improves health.

Simon Melov notes that the effects of antioxidant pills are fairly weak, since most of the nutrients don't get inside the cells where the free radical damage is occurring. Jay Olshansky dismisses megadose vitamin supplements as "a way to make expensive urine." A recent review in the *Journal of Gerontology* concurs: "The considerable enthusiasm for the use of micronutrient, especially antioxidant, supplements as anti-aging treatments or as treatments for specific diseases of later life is not supported by the currently available scientific literature."[32] Nevertheless, many of us cover our bets by taking supplements anyway.

Eukarion, a privately held biotech start-up in Massachusetts, is working with Melov to test its novel small-molecule antioxidant compounds. Melov has genetically engineered mice that don't produce superoxide dismutase, the body's own oxygen-scavenging enzyme that protects mitochondria from free radicals.

Such mice typically die within a week after birth. They suffer from enlarged hearts, damaged livers, and a spongiform brain ailment that looks very much like mad cow disease. But when these mice are injected with compounds that mimic the effects of superoxide dismutase and catalase (a compound that transforms the free radical hydrogen peroxide into water), they live four times longer. Eukarion is currently testing one of its compounds as a topical application to heal skin damaged by radiation treatments. It plans to test further compounds as treatments for degenerative neurological diseases in human beings.

Another dietary path toward potential rejuvenation is pointed out by a recent study conducted by Bruce Ames's team at Berkeley. They found that lethargic old rats are perked up by acetyl-l-carnitine and alpha lipoic acid. Their mitochondrial function improves, they become more active, and their memories get better.[33] A company called Juvenon, cofounded by Ames, has formulated a combination of the supplements that it claims will "slow down the clock on your aging cells." Keep in mind that while it works in rats, it may not work in people.

Researchers have long noted that as our bodies age, the levels of several key hormones begin to decline. Thus, hormone replacement therapy is second only to vitamin supplements in popularity among those seeking the fountain of youth. Its aim is to restore to their youthful level the hormones that decline with age. The most popular hormones eagerly purchased by would-be Methuselahs are DHEA, human growth hormone, melatonin, testosterone, and estrogen.

The notion that hormones could restore youth has a long and disreputable history. In the late nineteenth century the noted French physiologist Charles Edouard Brown-Sequard injected himself and his patients with extracts from the testicles of young dogs and guinea pigs, then declared that the treatments had restored his physical vigor and mental acuity. In the early twentieth century the American John Brinkley claimed to restore men's vitality by transplanting goat testicles into them. He performed more than sixteen thousand transplants before he died, though his medical license was revoked after some of his customers claimed a new compulsion "to chew sprouts."[34]

More recently, hormone replacement therapies took off after endocrinologist Daniel Rudman reported in 1990 that a dozen older men he had injected with growth hormone three times a week had increased their muscle mass, lost fat, tightened their skin, and lowered their cholesterol levels. Subsequent studies have shown that these quality-of-life benefits are real, but slight. In fact, regular exercise is more effective in obtaining most of the gains achieved from injecting growth hormone.

Furthermore, there is no evidence that growth hormone treatments increase longevity. Indeed, mice that overproduce growth hormone die sooner than normal mice, and fruit flies that underproduce growth hormone live longer than normal flies. In addition, some researchers suspect that supplementary growth hormone may increase the risk of cancer.

DHEA is another popular hormone consumed for antiaging purposes. It's the most abundant steroid in the body, yet nobody knows much about what it does. DHEA levels peak in a person's early twenties and decline thereafter. Interestingly, feeding DHEA to mice, which produce very small quantities of this hormone naturally, increases their life spans by 40 percent. In *Why We Age*, Austad notes that in the few scientifically valid human trials involving DHEA supplementation, the hormone produced "some improvement in immune response, muscle strength, and sleep patterns among the elderly."[35] Still, not much is known about its long-term effects, so most researchers advise caution.

Much has also been made of the so-called melatonin miracle. But rigorous testing of melatonin's effects on human beings has not yet been done. Mice that are fed melatonin live 5 percent longer but are at a greater risk of developing tumors. Again, most researchers advise caution.

So far, epidemiological evidence suggests that supplemental estrogen after menopause helps prevent osteoporosis. But recent research suggests that a combination therapy of estrogen and progestin actually increases a woman's risk of heart disease and breast cancer. Estrogen supplementation slightly increases the risk of ovarian cancer and promotes the growth of existing breast tumors. Using estrogen alone may delay the onset of certain diseases that become more common as women grow older, but there is no evidence that it increases users' life spans.

Testosterone levels generally drop in men as they age. Research on testosterone has lagged behind estrogen research, perhaps because of the unsavory treatments of the past and perhaps because of the bad reputation arising from steroid abuse among athletes. There are some indications that testosterone replacement can provide benefits to older men, including increasing muscle tone, overcoming erectile dysfunction, and improving overall sense of well-being. On the other hand, it might promote the growth of preexisting prostate tumors. There are other unwelcome side effects, including increased hairiness and acne. And there is no evidence that testosterone supplementation will increase longevity, either.

THE GENETIC IMPERATIVE

The final solution to aging may come not from diet—either restriction of calories or consumption of antiaging substances—but from genetics. Thomas Perls of Harvard University, for example, is hunting for longevity genes. Perls noticed a decade ago that people who live to be one hundred are often in remarkably good shape. Today, he runs the New England Centenarian Project, whose participants must be at least one hundred years old and have siblings who lived to be over ninety. By looking at DNA taken from some six hundred participants so far, Perls has found that a region on chromosome 4 appears to help its carriers become healthy geezers. The gene appears to have some effect on the regulation of fat metabolism.

Perls and his colleagues formed a company called Centagenetix to narrow the search for the longevity gene. Centagenetix has recently merged with Elixir Pharmaceuticals, which continues Perls's search for a Methuselah pill that mimics the activity of the proteins made by the longevity gene.

Looking further down the road, once the genes that promote disease (e.g., Alzheimer's, diabetes, cardiovascular problems) and those that promote longevity are identified, it will become possible for parents to select favorable genes for their progeny (see chapter 4). Already, nearly fifteen hundred healthy children have been born after their parents used preimplantation genetic diagnosis to select among eight-cell embryos to find the ones that were free of disease genes. In the future, parents might also select embryos that bear longevity-promoting genes and choose to implant those. Further in the future, parents will be able to add genes that improve their progeny's immune systems, mental acuity, and athletic abilities, perhaps by installing artificial chromosomes.

The Cambridge gerontologist Aubrey de Grey wants to genetically engineer mitochondrial genes into the nuclei of cells, where they would be better protected from the ravages of free radicals. He believes that once those genes are better protected they will not be so quickly mutated into the free radical death spiral. Once the vicious circle of mitochondrial mutations producing ever more free radicals is broken, longer life should result, he argues.

An even more visionary approach is being devised by Shaharyar Khan and Rafal Smigrodzki, two young researchers at the University of Virginia. They have made what may well be one of the great biomedical breakthroughs of the twenty-first century: a technique to replace old, worn-out mitochondria with fresh, new, revved-up ones. They call their technique protofection. Khan has established a startup company called Gencia to develop this research. In a brilliant tour de force, the Gencia researchers have devised a way to deliver the entire human mitochondrial genome (mitosome) into the mitochondria of living cells throughout an animal's body.

They create a protein-DNA complex designed to easily enter and specifically target mitochondria, by encapsulating healthy versions of the genes that make up the mitochondrial genome in mitochondrial transcription factor A (TFAM). Next they attach a protein transduction domain (PTD). (These are molecules that cells allow to easily cross through their protective membranes.) Finally, they attach what they call a mitochondria localization signal (MTD), which consists of malate dehydrogenase, a compound that mitochondria copiously guzzle as a normal part of their energy-producing cycle. Thus Khan and his colleagues have made a vector that delivers complete healthy mitochondrial genomes by passing swiftly and easily through cellular and mitochondrial membranes without provoking the body's immune response. Their vectors are like guided missiles, flying under the body's radar, able to deliver a payload safely to only one specific type of target.

The next step is even more amazing. They've created a way to eliminate old, damaged mitochondria and completely replace them with healthy new ones. First the healthy mitosomes are modified so that they do not express a particular DNA sequence present in normal mitochondrial genomes. (This change has no effect on the function of mitochondrial genomes.) They then add a gene for a restriction endonuclease to their healthy genome, which produces a protein that cuts up and disposes of normal mitochondrial genomes. In cell culture, new, healthy mitochondrial genomes completely replace old ones. Even more recently, they have used protofection in mice and found that fresh mitochondrial DNA had been taken up and activated in nearly all tissues. They conclude, "Protofection may be the first step towards therapeutic mtDNA replacement."[36]

Smigrodzki points out that misfiring mitochondria are thought to be a significant cause of Alzheimer's and Parkinson's diseases. Thus, he foresees first using tuned-up mitochondria to treat those diseases. To the extent that aging is the result of mitochondria misfiring and the production of damaging free radicals, this technique should rejuvenate cells and thus whole bodies. Smigrodzki predicts that mitochondria replacement will be used in therapies for neurodegenerative diseases in ten years or so and that antiaging whole-body "mito flushes" might be available in less than twenty years.

So what might a plausible antiaging regimen for the near future—say, twenty years—look like? First, people would want to minimize the damage caused by free radicals. Thus, they would take potions, pills, or shots of new, more effective antioxidants, similar to those being developed by the Buck Institute and Eukarion. Simultaneously, people seeking to live longer would take new sirtuin activators to fool the body into thinking that it is on the verge of starving, thus revving up the DNA repair mechanisms and reducing the amount of food stored as fat.

Nevertheless, damage to cells and tissues will inevitably accumulate, so perhaps antiaging specialists will treat patients with whole-body mito flushes every twenty years or so, thus replacing damaged mitochondria with new, healthy versions that produce fewer harmful free radicals.

De Grey offers a scenario in which efforts to achieve radical life extension reach "actuarial escape velocity (AEV)." Recall that for the last 160 years, average life expectancy has increased by two and one-half years per decade. What if increases in life expectancy rose at a rate of ten years or more per decade? "The escape velocity cusp is closer than you might guess," claims de Grey. "Since we are already so long lived, even a thirty percent increase in healthy life span will give the first beneficiaries of rejuvenation therapies another twenty years—an eternity in science—to benefit from second-generation therapies that would give another thirty percent, and so on ad infinitum."[37]

De Grey thinks that vast societal changes in lifestyle and expenditures would occur essentially overnight if people now living came to believe that extreme life extension technologies might arrive soon enough to benefit them. He suggests that these changes might be sparked if an experimenter were able to rejuvenate reliably a middle-aged mouse so that it lived three times longer than it would have normally. In fact, the Methuselah Mouse Foundation, founded by de Grey and others, is sponsoring a million-dollar prize with the aim of encouraging researchers to achieve this goal sooner rather than later.

De Grey concludes, "Once AEV is achieved, there will be no going back; rejuvenation research will be intense forever thereafter and will anticipate and remedy the life-threatening degenerative changes appearing at newly achieved ages with ever-increasing efficacy and lead time."[38]

NANOMEDICAL INSURANCE

But this focus on genetic and biotech interventions may be wrongheaded. After all, some argue, we don't fly because we sprouted wings, so neither will we live longer because we've fiddled with our genomes. Why not make machines that hunt down harmful disease organisms and repair damaged cells? That is the ambitious aim of nanomedicine.

Nanotechnology is the science and technology of building devices using single atoms and molecules. A nanometer is a billionth of a meter, a length just over the diameter of many atoms. Conceptually, nanotechnology and biotechnology are not all that distinct. In the words of Rita Colwell, the director of the National Science Foundation, "Life itself is nanotechnology that works."[39]

Proponents of medical nanotechnology—such as Ralph Merkle, a former research scientist at Xerox's Palo Alto Research Center and now a professor at Georgia Tech—outline an ambitious vision. "Nanotechnology will let us build fleets of computer-controlled molecular tools much smaller than a human cell and with the accuracy and precision of drug molecules," Merkle declared in an article on nanotechnology and medicine. He added, "These machines could remove obstructions in the circulatory system, kill cancer cells or take over the function of subcellular organelles."[40] Robert Freitas, author of the 1999 book *Nanomedicine*,[41] foresees a day when oxygen-carrying red blood cells could be supplemented by artificial respirocytes.

Freitas has already designed respirocytes composed of 18 billion precisely arranged atoms consisting of a shell of sapphire with an onboard computer and embedded with rotors to sort oxygen from carbon dioxide molecules. His respirocytes would be able to hold oxygen at one hundred thousand atmospheres of pressure. Just 5 ccs of respirocytes, one-thousandth the volume of the body's 30 trillion oxygen- and carbon-dioxide-carrying red blood cells, could supply enough oxygen to keep a person whose heart had stopped alive for four hours.

Freitas has also designed nanosubmarines consisting of 400 billion atoms that he calls vasculocytes. Ten million vasculocytes could fit on one square centimeter of blood vessel surface. Vasculocytes could inspect, repair, and recondition a person's entire vascular system, for example, removing atherosclerotic plaques before they burst to cause heart attacks or strokes. Perhaps Freitas's most ambitious medical nanorobots are microbivores that could supplement or replace infection-fighting white blood cells. Freitas claims his microbivores would be able to eliminate invading microbes one hundred times faster than the body's natural neutrophils do today.

If respirocytes, vasculocytes, and microbivores aren't wild enough, Freitas proposes a scheme that would replace the entire human circulatory system with a sapphire vasculoid weighing two kilograms. No heart, no blood—just a system of nanotech machines that would ferry oxygen, carbon dioxide, nutrients, and immune protective machines throughout your body, all encased in nearly unbreakable sapphire that would line your old-fashioned veins and arteries. Since 80 percent of what kills most people can be traced to the circulatory system—heart attacks, strokes, wounding, metastasizing cancer—such a vasculoid would dramatically increase one's life span. Freitas thinks the first models will be available in forty years.

"With nanotechnology we could someday be rebuilding our own bodies, regenerating organs, slowing down aging," predicted a bullish Samuel Stupp, professor of materials science and medicine at Northwestern University, at a National Science Foundation conference in 2002.[42]

Pharmaceutical manufacturers are already building new drugs—atom by atom, essentially—to treat diseases. Merkle believes that it will be another twenty to thirty years before the visionary technologies he foresees will be available. In the medical arena, nanotechnology and biotechnology may well be destined to meld together.

Nanotechnology also plays a role in what some consider the second-best alternative to living forever: cryonics. Cryonicists freeze people in liquid nitrogen with the hope that future technologies will be sufficiently advanced that the patients can be thawed out, revived, and cured of whatever ailments—including old age—afflicted them before they entered the deep freeze.

One problem facing cryonics enthusiasts is that no animal larger than a microscopic human embryo or a tiny tardigrade—an insect that measures only a couple hundred microns across—has yet been frozen and successfully revived. Freezing causes water in cells to expand, which disrupts them. But some researchers have developed a preservation technique called vitrification, which essentially glassifies cells.[43] This approach, it is claimed, causes far less disruption to internal cellular organization and destruction of cell walls. Scores of people have chosen to have either their full bodies or just their heads cryonically "suspended."

When it comes time to revive patients, the plan goes, nanotech machines will race through the patients' bodies, repairing the damage they have suffered from disease and freezing. Will it work? Who knows? Cryonicists put it this way: "The clinical trials are in progress. Come back in a century and we'll give you a reliable answer." Cryonicists divide the world into two groups: those who are experimenting with cryonics by being frozen, and those who just die and are buried. Which would you rather be in, they ask: the control group or the experimental group?

THE BATTLE OVER LIFE AND DEATH

What if a biomedical researcher discovered that our lives were being cut short because every human being was infected in the womb by a disease organism that eventually wears down the human immune system's ability to protect us? Until that discovery, the "natural" average life span was the biblical three score and ten years.

Once the discovery is made, another brilliant researcher devises a "vaccine" that kills off the disease organism. Suddenly, the average life span doubles to seven score (140 years). In a sense, this is exactly where we find ourselves today. There are no "vaccines" yet to cure the disease of aging, but biomedical researchers understand more with each passing year about the processes that cause the increasing physical and mental debilities that we define as aging. Aging is no more or less natural than cholera, smallpox, diabetes, arteriosclerosis, or any disease that cuts short human life.

Nevertheless, a number of prominent bioethicists and other policy intellectuals are arguing that we should oppose any such life-doubling "vaccine" on the grounds that it would interfere with the natural course of human life. Bioconservative opposition to attempts to radically extend healthy human life spans extends across the contemporary political spectrum. Liberal bioethicist Daniel Callahan makes three representative arguments against life extension. First, he points out that the "problems of war, poverty, environment, job creation, and social and familial violence" would not "be solved by everyone living a much longer life." Second, he asserts that longer lives will lead mostly to more golf games, not new social energy. "I don't believe that if you give most people longer lives, even in better health, they are going to find new opportunities and new initiatives," Callahan writes.

And thirdly, Callahan is worried about what longer lives would do to child bearing and rearing, Social Security, and Medicare. He demands that "each one of the problems I mentioned has to be solved in advance. The dumbest thing for us to do would be to wander into this new world and say, 'We'll deal with the problems as they come along.'"[44]

Callahan's first argument is a non sequitur. People already engage in lots of activities that do not aim directly at "solving" war, poverty, environmental problems, job creation, and the rest. Surely we can't stop everything until we've ended war, poverty, and familial violence. Anti-aging biomedical research obviously wouldn't exacerbate any of the problems listed by Callahan and might actually moderate some of them. If people knew that they were likely to enjoy many more healthy years, they might be more inclined to longer-term thinking aimed at remedying some of the problems he frets about.

Second, Callahan's "longer life equals more golf" argument is not only condescending, it ignores the ravages that physical decline visits on people. Callahan, age seventy-three, sees a lack of "new energy" among his confreres. Even if people are healthy at age seventy-five, their "energy" levels will be lower than at age thirty. They may not begin "new initiatives" because they can't expect to live to see them come to fruition.

But diminishing physical energy isn't the only problem; there is also waning psychic energy. "There's a factor that has nothing to do with physical energy. That is the boredom and repetition of life," he argues. "I ran an organization for 27 years. I didn't get physically tired. I just got bored doing the same thing repetitiously."[45]

It doesn't seem reasonable to conclude that, just because Callahan is bored with his life, we all will become so. Modern material and intellectual abundance is offering a way out of the lives of quiet desperation suffered by so many of our impoverished ancestors. The twenty-first century will offer an ever-increasing menu of life plans and choices. Surely exhausting the coming possibilities will take more than one standard lifetime to achieve. Besides, if you do want to play endless games of golf and can afford it, why is that immoral? And if you become bored with life and golf, well, no one is making you hang around.

Doubling healthy human life expectancy would create some novel social problems, to be sure, but would they really be so hard to deal with? Callahan cites the hoary example of brain-dead old professors blocking the progress of vibrant young researchers by grimly holding onto tenure. That seems more of a problem for medieval holdovers like universities than for modern social institutions like corporations.

Assuming it turns out that, even with healthy, long-lived oldsters, there is an advantage in turnover in top management, then corporations and other institutions that adopt that model will thrive and those that do not will be outcompeted. Besides, even today youngsters don't simply wait around for their elders to die. They go out and found their own companies and other institutions. Bill Gates didn't wait to take over IBM; he founded Microsoft at age twenty. Nor did human genome sequencer J. Craig Venter loiter about until the top slot at the National Institutes of Health opened up. Sergey Brin and Larry Page founded Google, currently the world's most popular Internet search engine company, in their twenties. Michael Dell founded Dell Computers, the world's largest seller of personal computers, at age nineteen. In politics, we in the United States already term limit the presidency, as well as many state and local offices. And empirical evidence cuts against Callahan's worries that healthy geezers will slow economic and social progress. After all, social and technological innovation has in fact been most rapid in those societies with the highest average life expectancies. Yale economist William Nordhaus argues that the huge increase in average life expectancy from forty-seven years to seventy-seven years since 1900 today has been responsible for about half the increase in our standard of living in the United States.[46]

Callahan's failure of imagination when it comes to public policy conundrums such as Social Security and Medicare is breathtaking. Folks will be chronologically older, but not elderly in the current sense. Thus, the standard age when those payoffs begin will obviously rise, as the chronologically aging but not physically older will be expected to continue to be productive and support themselves. Assuming that age-retardation is possible, the many illnesses and debilities that accompany aging will be postponed. If one is going to live to be 140, one has a lot of time to plan and save for the future. Consider that $300 compounds to $500 million in just 150 years. Think, too, of the trillions of dollars that would be freed up for innovative projects and products if the obligations for retirement and old-age medical benefits no longer needed to be met. Few would choose a Social Security check over sixty-five additional years of healthy life.

And his assumption of a child crisis in a world of long-lived people seems based on the idea that healthy oldsters would be less interested in reproducing. A first response might be: So what? Shouldn't the decision to have children be up to individuals? After all, already countries with the highest life expectancies have the lowest levels of fertility.

This lack of interest in progeny would have the happy side effect of addressing the concern that doubling human life spans might lead to overpopulation. No one can know for sure, but it could well be that bearing and rearing children would eventually interest long-lived oldsters who would come to feel that they had the time, the life experiences, and the resources to do it right.

Callahan's final demand—that all problems that doubled healthy life spans might cause, be solved in advance—is just silly. Humanity did not solve all of the problems caused by the introduction of farming, electricity, automobiles, antibiotics, sanitation, and computers in advance. We proceeded by trial and error and corrected problems as they arose. We should be allowed do the same thing with any new age-retardation techniques that biomedical research may develop. And we'll be happy to do so.

Callahan is far from alone in favoring mortality. The prospect of physical immortality fills left-wing environmentalist Bill McKibben "with blackest foreboding." He argues, "It would represent, finally, the ultimate and irrevocable divorce between ourselves and everything else. The divorce, first of all, between us and the rest of creation."[47]

McKibben would do better to ask why we would want to stay married to Nature anyway. She has certainly been an inconstant spouse, liberally afflicting us with nasty surprises such as birth defects, diseases, earthquakes, hurricanes, and famines. An amicable separation might be good for both Nature and humanity. The less we depend on Nature for our sustenance, the less harm we do her.

Setting that aside, why does McKibben believe that death is good for us? "Without mortality, no *time*," writes McKibben. "All moments would be equal; the deep, sad, human wisdom of Ecclesiastes would vanish. If for everything there is an endless season, then there is also no right season. The future stretches before you endlessly flat."[48]

Actually, the deep sad wisdom of Ecclesiastes is a very powerful human response of existential dread to the oblivion that stretches endlessly before the dead. "For the living know that they shall die: but the dead know not any thing, neither have they any more a reward; for the memory of them is forgotten. Also their love, and their hatred, and their envy, is now perished; neither have they any more a portion for ever in any thing that is done under the sun," writes the preacher (Eccles. 9:5–6). If death is not inevitable, humanity will have the opportunity to learn happier new teachings and glad new wisdom.

If the endless future turns out to be as horrible as McKibben imagines it to be, then people will undoubtedly choose to give up their empty, meaningless lives. On the other hand, if people opt to live yet longer, wouldn't that mean they had found sufficient pleasure, joy, love, and even meaning to keep them going? McKibben's right: We don't know what immortality would be like. But should that happy choice become available, we can still decide whether or not we want to enjoy it. Even if the ultimate goal of the biotechnological quest is immortality, what will be immediately available is only longevity. The experience of longer lives will give humanity an opportunity to see how it works out. If immortality is a problem, it is a self-correcting problem. Death always remains an option.

It is not only leftists who object to the quest for increased longevity. One of the more prominent conservative opponents of this sort of progress is Francis Fukuyama. Disturbingly, when asked if the government has a right to tell its citizens that they have to die, Fukuyama answered, "Yes, absolutely."[49]

Among other things, Fukuyama fears that our efforts to extend human life will create a nursing home world, filled with aging, miserable, debilitated people draining resources from the young to keep themselves alive. Leon Kass also points to the "undesired consequences of medical success in sustaining life, as more and more people are kept alive by artificial means in greatly debilitated and degraded conditions."[50] Here both are engaged in what might be called Struldbruggism, after an episode in *Gulliver's Travels* in which Gulliver visits the land of the Luggnuggians, among whom are occasionally born immortals called Struldbruggs. This is no blessing, since the immortals still grow ever weaker, sicker, and demented.

As a justification for his fear of a nursing home planet, Fukuyama points out that as many as 50 percent of those who reach age eighty-five may have Alzheimer's disease. Of course, this disease is terrible, but keep in mind that, according to the National Institutes of Health, Alzheimer's "is not a normal part of aging."[51]

Our already successful efforts to lengthen life *have* led to an increase in Alzheimer's disease, because people are now living long enough to get the disease. Yes, Fukuyama has medical science dead to rights: Our efforts to lengthen life by preventing or curing disease *have* been spectacularly successful. Statisticians at the National Institutes of Health "calculate that if death rates were the same as those of 30 years ago, 815,000 more Americans a year would be dying of heart disease and 250,000 more of strokes."[52] As horrible as Alzheimer's disease is, the fact that millions of people now live long enough to get it represents a real benefit: the additional years of healthy life they had in order to get old enough to develop the disease in the first place. As even Callahan has acknowledged: "Evidence has shown that as average life expectancy has increased, the disabilities traditionally associated with aging have actually decreased."[53]

The cutting edge of medical life-advancing technology slices through Fukuyama's nursing home fears by aiming directly at preventing aging, not just ameliorating the diseases—such as cancer, Alzheimer's, and heart disease—that often accompany it. The goal of research on aging is not to turn us into a race of miserable Struldbruggs. As Olshansky puts it, "We don't want to make ourselves older longer, we want to make ourselves younger longer."[54] Characteristically, Fukuyama and Kass ignore the real goals of antiaging research, misleading the public with a gruesome scenario in hopes of frightening them into embracing their pro-death dogma.

Fukuyama argues that *my* life extension might be a problem for all the rest of us. He claims that "life extension seems to me a perfect example of something that is a negative externality, meaning that it is individually rational and desirable but it has costs for society that can be negative."[55]

Living longer as a negative externality to society? For a man who has a very sophisticated understanding of political philosophy, this argument is surpassingly strange. After all, philosopher Thomas Hobbes made the point in his *Leviathan* that individuals form society for the purpose of making sure that their lives are not "solitary, poor, nasty, brutish, and short." In other words, individuals do not exist for society—society exists for the happiness of individuals. And who doubts that most people would be happy to live longer, healthier lives?

Fukuyama needs to keep in mind that past generations have already chosen to more than double average human life expectancy. In 1900 average global life expectancy was around thirty-two years; today it is sixty-seven years. Our grandparents and parents did not ask our permission to double global life expectancy; they just went ahead and did it. It is unlikely our descendents will have any more reason to regret our decision than we have to regret that of our forebears.

Fukuyama's final concern echoes Callahan's worry about eternal games of golf: He worries that people will be "eager to continue his or her mediocre life for as long as possible without worrying about some of these higher questions about what life is used for."[56] Kass agrees: "The pursuit of perfect bodies and further life extension will deflect us from realizing more fully the aspirations to which our lives naturally point, from living well rather than merely staying alive."[57]

Setting aside Fukuyama's and Kass's harsh character judgments of their fellow human beings, they offer a false dichotomy. To live well, one must first *be* alive. I seriously doubt that people granted longer lives will fritter away their extra time watching reruns of *Gilligan's Island* (though some might, and it would be their business). Instead, they may well engage in longer-run projects such as ecological restoration or space exploration.

The most politically significant statement of opposition to longevity research came from the President's Council on Bioethics's report *Beyond Therapy.* It deals comprehensively with the perceived problems with radical life extension in a chapter titled "Ageless Bodies." Nearly all of the concerns in *Beyond Therapy* reflect the particular longstanding anxieties of Kass, the panel's chairman. He raised these issues publicly as early as 1971, when he served as the executive secretary of the National Academy of Sciences' Committee on Life Sciences and Social Policy. For example, an article in the February 23, 1971, *New York Times* quotes him predicting, "If efforts to understand and alter the aging process are pursued full tilt, by the end of this century the 'useful' life span of some individuals may be extended 20 to 40 years."[58]

Was this cause for celebration? Certainly not. The *Times* story goes on: "The young generation of today, he asserted, is the first to grow up without fear of lethal diseases. Hence its members, as a whole, have neither faced death as a real threat to themselves nor seen it take a certain number of their contemporaries, he said. Dr. Kass argued that people were therefore becoming less and less conditioned to accepting death as a natural and merciful way to terminate the aging process."[59] Indeed we are.

Unfortunately, Kass was wrong about how much biomedical progress would be able to boost average life expectancy by the end of the century. In 1971 average US life expectancy was seventy-one years; today it has increased to more than seventy-seven years—up six years, but a far cry from twenty to forty years. But it is clear that for more than thirty years now, Kass has been doggedly pursuing his pro-death agenda.

Beyond Therapy also recapitulates many of the concerns expressed by Callahan and Fukuyama. For example, the chapter titled "Ageless Bodies" suggests that much longer lives would weaken our "commitment and engagement." Now we live with the knowledge that we will soon die and thus "the more likely we are to aspire to spend our lives in the ways we deem most important and vital."[60] Second, our "aspiration and urgency" might flag because we would ask, "Why not leave for tomorrow what you might do today, if there are endless tomorrows before you?"[61]

Third, the chapter describes worries about our commitment to "renewal and children," wondering whether "a world of men and women who do hear the biological clock ticking and do not feel the approach of their own decline might have far less interest in bearing—and more important, caring for—children."[62] Fourth, our "attitudes toward death and mortality" might shift dramatically because "an individual committed to the scientific struggle against aging and decline would be less prepared for death, and less accepting of death, and the least willing to acknowledge its inevitability."[63] And fifth, age retardation might "leave the individual somewhat unhinged from the life cycle" such that we would not be able "to make sense of what time, age, and change should mean to us."[64]

In addition, "Ageless Bodies" highlights three areas of societal concern. Significant age retardation would disrupt the succession of "generations and families." This succession "could be obstructed by a glut of the able,"[65] the chapter suggests, since cohorts of healthy geezers would have no intention of shuffling off this mortal coil to be replaced by younger people. Longer lives could also slow down "innovation and change" since "serious innovation, and even successful adaptation to change, is . . . often the function of a new generation of leaders."[66] Finally, even if we are not aging individually, we will need to worry about "the aging of society" that would then result. Societies composed of people whose bodies do not age significantly might "experience their own sort of senescence—a hardening of the vital social pathways."[67]

Despite all of the concerns cited by the President's Council on Bioethics, *Beyond Therapy* tellingly goes on to admit that "*powerful as some of these concerns are, however, from the point of view of the individual considered in isolation, the advantages of age-retardation may well be deemed to outweigh the dangers*.[68] Indeed.

As we've seen, there are many rejoinders to Kass's and the council's concerns about boosting healthy human life spans, but council member Michael Sandel, a professor of political philosophy at Harvard, deftly slices through the Gordian knot of issues tangled together in the report, noting simply that if longer life spans are bad, then shorter ones must be good:

> Are the background conditions in human self-understandings for the virtues just about right now at 78 years . . . or such that they would be eroded and diminished if we extend it to 120 or 150, or 180? Is it the suggestion that back when [average life expectancy] was 48, rather than 78, a century ago . . . that the virtues we prize were on greater display or more available to us? And if so, would that be reason to aim for, or at least to wish for or long for, a shorter life span, rather than a longer one?[69]

Elsewhere, Kass has argued that "immortality is a kind of oblivion—like death itself."[70] He adds, "Mortality as such is not our defect, nor is bodily immortality our goal. Rather mortality is at most a pointer, a derivative manifestation, or an accompaniment of some deeper deficiency . . . the human soul yearns for, longs for, aspires to some condition, some state, some goal toward which our earthly activities are directed but which cannot be attained in earthly life. . . . Man longs not so much for deathlessness as for wholeness, wisdom, goodness, and godliness."[71]

Taking Kass on his own terms, what can he mean by asserting that immortality would be a kind of oblivion? Keep in mind: *Whatever the risks of "oblivion" posed by immortality, they must surely be balanced against the certain oblivion of the grave.*

Kass points to the frivolity of the lives of the gods in Greek myths. Nothing is serious to them because eternity stretches endlessly before them. Their choices are whimsical and capricious. They are not, strictly speaking, moral beings, because no decisions or choices are final—among themselves the passage of time can reverse all harms and all insults. It's all just a game and mortals are of no more real significance to them than mosquitoes are to us.[72]

Do these myths provide a warning to heedless humanity about the pursuit of immortality? Perhaps not. After all, even an immortal being can decide to become mortal. For example, according to Christianity, God himself became a man, and he died. To the extent that the oblivion of death gives urgency and poignancy to life, it will still do so even for very long-lived people. It is noble now to give one's life to save others, but how much more noble it will be when defenders sacrifice centuries rather than decades of their lives for others?

Another problem with the Greek gods as cautionary models is that they existed in an eternity with no progress, no goals other than playing the same endless round of games among themselves. Many visionaries who argue for dramatically extending human life spans foresee a day when people can use superadvanced technologies to become very much like gods, with vast new abilities to apprehend new knowledge and to achieve new goals. But for a very long time, such godlike people will clearly be neither omniscient nor omnipotent—so plenty of adventure and wisdom will remain to seek after in the wider universe.

What about Kass's claim that "man longs not so much for deathlessness as for wholeness, wisdom, goodness, and godliness"? This is again simply a false dichotomy. One can long for deathlessness *and* wholeness, *and* wisdom, *and* goodness, *and* godliness. Or is Kass saying that we seek these virtues only because we are mortal? If so, it cannot be that they are somehow intrinsically or transcendently important. In any case, longer lives allow us to seek further those other human virtues Kass says he prizes. Defeating death is not the final act in our technological or spiritual quest—it is merely a vital prelude to our further pursuit of wholeness, wisdom, goodness, and, yes, even godliness. Surely, for all too many people, perhaps for most of us, the pursuit of these virtues has been or will be cut short by death well before we have managed to attain them.

Prominent theologians are at odds over Kass's fears that longevity conflicts with godliness. The Extended Life/Eternal Life Conference sponsored by the University of Pennsylvania and the John Templeton Foundation brought together a number of theologians to consider the morality of trying to radically extend human life spans. Christian perspectives were offered by theologians Diogenes Allen from Princeton University and Richard John Neuhaus of the Institute of Religion and Public Life. In general the theologians see death as a blessing because (1) this life cannot satisfy our longing for perfect love, (2) indefinite continuation would be dreary, (3) the very shortness of life calls for self-examination, and (4) it breaks through our solipsistic self-regard.[73]

Interestingly, Allen pointed out that a belief in justice has been a powerful motivation for people to postulate an afterlife. Good people suffer in this life, but will be rewarded in the next. I can't resist suggesting that with radical life extension perhaps more people will live to see justice done in this life. Philosopher Eleonore Stump of St. Louis University suggested that medieval Christians would have thought that our efforts to postpone death "was a perplexing stupidity on our part" since it prolonged our separation from God.[74]

However, it was generally agreed by the panel that there was no absolute theological objection to extending human life. After all, Methuselah lived 969 years according to the Bible. While, the Christian theologians were ambivalent about increasing healthy human life spans, the Jewish perspective was inspiring. Rabbi Neil Gillman was refreshingly blunt: "There is nothing redemptive about death. Death is incoherent. Death is absurd."[75] In Judaism the primary metaphor for God is that God is Life, he explained. The best moment came during the question-and-answer session, when Kass asked the rabbi if Jewish tradition would endorse prolonging human life for twenty years? Yes, answered the rabbi. Forty years? Yes. One hundred years? Yes. The indefinite prolongation of life is a moral good, then? "Yes, yes, yes," answered Rabbi Gillman.

Charles Harper, the director of the Templeton Foundation, noted,

"Stasis is not an option. The world is fundamentally out of control and there is a free competition of visions in the context of constant change." He urged the religious community to stop nostalgically hoping for stasis and instead build strong visions of the good and the virtuous that can compete and inspire human creativity and the will to resist evil. "I favor radical life extension," said Harper. "So I say hooray for life and hooray for more of it."[76] Amen.

In the end, *Beyond Therapy* acknowledges that it is a "reasonable expectation" that "if effective age-retardation technologies become available and relatively painless and inexpensive, the vast majority of us would surely opt to use them, and they would quickly become popular and widely employed."[77]

Ultimately, the promortalists on the right and the left may significantly delay the development of longevity treatments, but the chances that they can stop them forever seem small. "It's too alluring," says Olshansky. "It's been the dream of humanity forever. How can we not?"[78] Austad agrees. "People want this so badly, it's going to happen no matter what the government does," he predicts. "The government can help or it can hinder this research, but it will happen."[79]

"A dramatic increase in life span is inevitable," predicts de Grey "We understand aging at the molecular level sufficiently to not just imagine interventions to retard aging, but enough that we can describe them. It's an engineering project now, not a scientific one. We just don't know how long it will take."[80]

What Kass, Callahan, Fukuyama, McKibben, and other bioconservatives forget is that the effort to extend human life spans is a perfect flourishing of our human nature. The highest expression of human dignity and human nature is to try to overcome the limitations imposed on us by our genes, our evolution, and our environment. Future generations will look back at the beginning of the twenty-first century with astonishment that some very well-meaning and intelligent people actually wanted to stop biomedical research just to protect their cramped and limited vision of human nature. They will look back, I predict, and thank us for making their world of longer, healthier lives possible.

CHAPTER 2

FINAL VICTORY OVER DISEASE?

Building Humanity's Extended Immune System

"I certainly am happy to still be in the land of the living," declares Bernis Teaters, a five-year survivor of lung cancer from Friendswood, Texas. "It's wonderful," agrees Alfredo Gonzalvo, another five-year lung cancer survivor. Both are alive today thanks to a pioneering anticancer gene therapy.[1] In 1999 researchers at the M. D. Anderson Cancer Center in Houston, Texas, injected their lung tumors with a vaccine of deactivated viruses containing the cancer-killing p53 gene while they underwent radiation treatments. The p53 gene in the anticancer vaccine instructed their tumor cells to commit suicide. As a result, the two are cancer free today. A similar vaccine is already available in China and will soon be approved for use in the United States.

An earlier beneficiary of a pioneering gene therapy is Cynthia Cutshall, a student at Kent State University. Getting through college took hard work, of course. But when her family and friends celebrate her graduation, it will also be thanks to a modern medical miracle. In 1990 Cynthia was a nine-year-old suffering from a life-threatening genetic disease, severe combined immunodeficiency (SCID), which completely disabled her immune system. Her body could not protect her against the assaults of bacteria and viruses. In the 1970s SCID had became widely known as "bubble boy disease," when the world learned of David Vetter, a Texas boy with SCID who lived for twelve years in a germ-free plastic bubble. Unfortunately, David lived before the age of gene therapy and he died from complications of an earlier treatment.

Like David Vetter, Cynthia was forced to live in a hyperclean environment, cut off from most contact with the outside world. She could neither play with other children nor attend school. Her illness was caused by a mutation in the gene that produces a crucial enzyme, adenosine deaminase (ADA), whose absence allows the buildup of substances toxic to immune system cells.

In 1990 Cynthia and another little girl from Ohio who suffered from SCID, four-year-old Ashanthi DeSilva, became the first patients in the world to receive therapies aimed at curing their disease by correcting their defective genes. Gene therapy introduces healthy genes into a patient's body to replace the damaged ones. To achieve this, a normal gene is isolated and put into a delivery vehicle called a vector. The vector, often made from a disabled virus, infects target cells in the malfunctioning tissues or organs. Once the vector reaches its goal, normal genes are released and, hopefully, begin to produce the correct proteins, restoring faulty cells to normal.

In the cases of Cynthia and Ashanthi, a team of researchers led by Michael Blaese at the National Cancer Institute in Bethesda, Maryland, aimed to repair the girls' defective ADA genes by infecting their own white blood cells with viruses modified to carry healthy versions of the ADA gene and then reinfusing the blood cells into the girls. It worked, at least somewhat. The girls' immune systems developed enough capacity to provide some protection against infection. This allowed them to lead more normal lives, joining their friends in school. Nevertheless, both still require injections of a drug called PEG-ADA to boost their immune systems.[2]

It was not until a decade later that the first complete cure of a genetic disease was achieved by replacing malfunctioning genes with healthy ones. In April 2000, French researchers led by Alain Fischer at Hospital

Necker in Paris announced that they had successfully treated three infants suffering from another version of SCID.[3] In this case, the researchers replaced the defective gene by removing bone marrow stem cells from the infants and infecting them with viruses modified to contain the healthy gene. The marrow stem cells incorporated the healthy genes and were infused into the infants, where the corrected cells took up residence in their bones. The good news is that the corrected cells grew and outcompeted the faulty ones. The French researchers have successfully used gene therapy to treat seven other children who suffered from SCID. All of the children developed functional immune systems and began to live normal lives. In 2002 it looked as though the hope for a cure might be short-lived when it was reported that the two youngest children had developed leukemia. In early 2005 French gene therapy researchers announced that a third child treated for SCID had come down with leukemia.

In 2003 and 2004 intensive research revealed that the corrected gene had been installed in the marrow stem cells in such a way that it turned on an oncogene (a gene that promotes the development of cancer) associated with leukemia. All three children received antileukemia chemotherapy. In the earlier two cases, one of the boys now appears to be in complete remission and is no longer being treated for leukemia; unfortunately, the other died of the disease. The third boy is responding well to antileukemia treatments. Keep in mind that without treatment, infants with SCID usually die before their first birthdays. Although the leukemia cases are disappointing, Mark A. Kay, a professor of pediatrics and genetics at Stanford University, notes, "Taking a disease that is pretty much fatal . . . if you can get a 60 or 70 percent cure rate, you have to balance that out."[4] Researchers have also discovered that the problem will not likely arise in gene therapies designed to treat other diseases and have even devised a way around the problem for future SCID patients.

When the initial leukemia cases were announced, regulatory authorities in the United States and France rushed to shut down all gene therapy clinical trials until someone figured out what had happened.

Previously, gene therapy trials had been shut down in the United States by the Food and Drug Administration (FDA) after the death of teenager Jesse Gelsinger in 1999.[5] Gelsinger suffered from a genetic disease called partial ornithine transcarbamylase (OTC) deficiency, in which his liver was unable to process proteins efficiently, leading to the toxic buildup of ammonia in his body. While most children die at a young age from the disease, Gelsinger was controlling his condition with medicine. In order to help researchers devise genetic treatments for children who suffered from worse versions of his disease, Gelsinger volunteered to participate in a gene therapy clinical trial at the University of Pennsylvania.

The Penn researchers used modified adenoviruses as vectors to replace the defective gene in his liver with healthy ones. Unfortunately, shortly after Gelsinger's liver was infused with the modified viruses, he experienced a massive immune system reaction to the vectors, which led to multiple organ system failure. He died four days after he was injected with the vectors. A subsequent inquiry found that the researchers, led by James Wilson, had violated the protocols of the clinical trial and had misled Gelsinger and bioethics overseers about the safety of the procedure.[6] In 2002 the Gelsinger family reached an out-of-court financial settlement in a wrongful death lawsuit. In 2005 the researchers and their associated institutions paid $1.1 million to the federal government to settle civil fraud charges. Wilson's Institute for Human Gene Therapy was shut down and he was forbidden to engage in human gene therapy trials in the future. Wilson returned to the University of Pennsylvania's Medical School, where he has continued to work on animal models of gene therapy.

Since 1990 only a few thousand patients have been treated with any forms of gene therapy, and successes are still disappointingly few. As with any new technology, there is generally a lot of hype in the early stages. Still, therapies based on replacing faulty genes with healthy ones are just the first halting baby steps of a revolution in biomedicine leading, if all goes well, to the creation of humanity's extended immune system.

The notion of an extended immune system is adapted from evolutionary theorist Richard Dawkins's idea of an extended phenotype. Dawkins observed that genes not only mold the bodies of organisms, but also shape their behaviors. Some of those behaviors result in the creation of inanimate objects—technologies, if you will—that help organisms to survive and reproduce, such as beaver dams and bird nests. Nothing could be more natural to human beings than striving to use our technological prowess to liberate ourselves from our biological constraints. Biotechnologists are now turning humanity's technological prowess from the outside world to the intimate inner worlds of our bodies, tissues, and cells. Our immune systems protect our bodies from foreign substances and pathogenic organisms by producing the immune response. Our immune systems also guard our bodies against cells, such as cancer cells, that betray the body. The immune response results from an integrated system of organs, tissues, cells, and cell products such as antibodies.

Thanks to the development of a whole suite of new diagnostics and therapies, human intelligence is extending our technological control to our immune systems, much as we have extended and enlarged the effects of our bodies in the world by means of our tools. Over the next several decades, biomedical technologies will enable us to better protect our bodies against foreign invaders and traitorous cells.

Progress is moving so rapidly and on so many fronts that we can only touch on a few of the fascinating breakthroughs that are enabling humanity to construct its extended immune system. These include gene transfer and replacement therapies, gene therapies aimed at preventing and curing cancer, medicines to prevent and cure infectious diseases, and new techniques to turn genes on and off as a way to defeat disease. We will also see how maintaining scientific openness and the development of a robust biotechnological capability are vital to protecting humanity against the threat of bioterrorism.

FIXING BROKEN GENES

We began this chapter by looking at the first successes and failures of gene transfer therapies involving cancer vaccines, SCID, and OTC. Clinicians have identified about four thousand diseases that result from single gene defects or mutations. Ideally, each of them could be cured if researchers could find a way to replace malfunctioning genes. Researchers have become proficient at curing genetic diseases such as hemophilia in lab mice and dogs,[7] blindness in dogs,[8] sickle cell disease in mice,[9] muscular dystrophy in mice,[10] and cystic fibrosis in mice.[11] However, translating these proof-of-concept animal therapies to people has been difficult, to say the least. For example, researchers at the University of Pennsylvania injected viruses carrying the healthy gene for blood clotting Factor IX, the lack of which causes hemophilia B, into the muscles of patients. Their muscles took up the healthy genes, but Factor IX did not increase to levels sufficient to prevent excessive bleeding.[12]

One of the chief problems bedeviling gene therapies based on viral vectors is that the human immune system treats them as it would any disease organism: It tries to destroy them. Furthermore, the viruses modified to deliver healthy genes are typically grown in mouse cells fed on cow serum, from which they pick up factors that human immune systems treat as foreign and attack. Nevertheless, scientists are improving methods to use viral vectors to deliver healthy genes.

To get around the problem of immune response provoked by vectors, researchers are developing new techniques for delivering healthy genes to fix faulty ones. Copernicus Therapeutics, based in Cleveland, Ohio, has devised a way to compact DNA onto nanoparticles that can slip through the pores in cell membranes. Copernicus is pioneering a way to use genes delivered as nanoparticles to treat cystic fibrosis. Cystic fibrosis is a genetic respiratory disease afflicting about thirty thousand people in the United States, with one thousand new cases diagnosed each year. With this disease, the gene for a chloride channel is impaired, allowing the buildup of thick mucus that clogs the lungs and blocks the ducts of the pancreas. People with cystic fibrosis often suffer life-threatening infections. Their average life expectancy is around thirty-one years.

Copernicus's nanoparticles are about the size of viruses, measuring less than twenty-five nanometers (.000000025 of a meter) in diameter. In clinical trials, these nanoparticles were sprayed into the nostrils of people with cystic fibrosis. Later tests showed that the genes were taken up and functioned properly with no noticeable side effects. New clinical trials will test whether these nanoparticles can safely and efficiently transfer a normal copy of the cystic fibrosis gene into epithelial cells that line the lungs.[13]

Gene transfer technologies are not used only to correct faulty genes. They can be used to jump-start healing processes, too. Researchers at St. Elizabeth's Medical Center and Tufts University in Boston reported very promising gene therapy research designed to repair hearts damaged by atherosclerosis. In clinical trials, the gene for vascular endothelial growth factor 2 (VEGF-2) was inserted via catheter directly into the blocked coronary arteries of patients suffering from angina. The idea is that the infused VEGF-2 genes will entice the patient's heart to start growing new blood vessels to go around the blocked ones, essentially growing their own heart bypasses without surgery. The patients chosen for the research had tried all other treatments for their heart disease, including heart bypasses and balloon angioplasty. Most could barely walk one hundred feet before experiencing crushing chest pain.

The researchers reported that in one trial, 70 percent of the patients had a significant decrease in angina symptoms. One patient, the Rev. Charles Wilson of Charlotte, North Carolina, explained that before the gene therapy, he suffered "[t]wenty to twenty-five severe angina episodes a day. Something as simple as shaving would bring on angina. Now, I'm back in the pulpit."[14]

In one particular protocol in which thirteen patients were treated with the VEGF-2 gene, nine had increased blood flow to damaged parts of their hearts. In another double-blind study of nineteen patients in which twelve received VEGF-2 and seven did not, eight of the twelve treated with VEGF-2 had increased blood flow to their hearts. In other words, the VEGF-2 gene therapy worked.[15]

GENE THERAPY ABUSE?

But even before gene transfer therapies are perfected for curing disease, some people are already worried that these techniques could be used for genetic enhancement. For example, perfected gene transfer techniques could be used to insert desired genes into fertilized eggs and developing embryos, either to correct genetic flaws or to provide genetic enhancements. When the child is born, the inserted genes will be incorporated in all her tissues and she will be able to pass them along to her future progeny (see chapter 5).

Gene transfer therapies could also be used to enhance adults. Lee Sweeney, chair of the Department of Physiology at the University of Pennsylvania, discovered in 1998 that he can insert the gene for IGF-1 into the muscles of mice and prevent the muscle wasting that occurs as a result of both aging and the genetic disease muscular dystrophy. Sweeney is conducting this research with the goal of treating muscular dystrophy and preventing people from losing their muscles as they age, a leading cause of fatal falls among the elderly.[16] In fact, Sweeney's old mice experienced a 19 percent increase in muscle mass. Even better, the older mice experienced a 27 percent boost in muscle strength, and both muscle mass and function were restored to their youthful levels.[17]

These early therapeutic successes in animals almost immediately provoked much gnashing of teeth over the possibility that athletes might begin to engage in "gene doping." Athletes would seek out physicians to inject IGF-1 genes into their muscles in order to boost their size and power. "We're trying to work with muscular dystrophy," says Sweeney. "But we're drawing a road map for how the athlete of the future could obtain tremendous performance enhancement. We need to be aware of what's possible so people can start to look for it."[18]

Before rushing out to inject performance-enhancing genes, athletes would be well advised to heed the results of experiments in which the EPO gene, which promotes the production of red blood cells, was injected into the legs of eight rhesus monkeys. It is well known that boosting red blood cells, which carry oxygen to the muscles, improves an athlete's stamina. Infamously, many East German Olympic athletes had red blood cells injected into them just before competition. In the EPO gene experiment, half of the monkeys *overproduced* red blood cells. This would have turned their blood to sludge, causing them to die of strokes, had they not had their blood thinned every two weeks. The other half of the monkeys mysteriously suffered from a fatal anemia in which the production of red blood cells was completely shut down.[19] With those cautions in mind, few doubt that such enhancements will certainly one day work. So would gene doping be cheating?

"I want to be sure when I cheer that I'm cheering for the [athlete] and not his or her chemist," sniffs Leon Kass.[20]

The plain fact of the matter is that athletes are already more genetically gifted than most of us, and some even more so than others. For example, Finnish cross country skier Eero Maentyranta won two gold medals at the 1964 Winter Olympics. Certainly he trained hard, but Maentyranta had an advantage: he was born with a variant of the EPO gene that caused him to produce 25 to 50 percent more red blood cells than the average person.[21]

In 2004 a five-year-old German boy with unusually large and strong muscles was found to have a mutation that deactivated a gene that would normally slow muscle growth. Interestingly, his mother, a competitive sprinter, was found to have one copy of the mutated gene herself.[22] Should he and his mother be forbidden to compete although they come by their advantages "naturally"? Interestingly, Wyeth Pharmaceuticals is now investigating ways to deactivate this same gene as a possible therapy to prevent muscle wasting in the elderly.

In August 2004 researcher Ronald Evans reported that his team at the Salk Institute had been able to genetically engineer mice to produce Type I muscle fibers that turns them into endurance runners, so-called Marathon Mice.[23] The Marathon Mice were able to run 92 percent longer and twice as far as normal mice.[24] It is now known that gene variants in human athletes account for the balance between Type I (endurance) muscle fibers and Type II (fast twitch, sprinter) muscle fibers. In fact, in November 2004 an Australian company began offering its $100 Sports Gene Test to help athletes decide which sports they should specialize in based on whether their genes code for more endurance or fast twitch muscle fibers.[25]

Of course, each sporting authority may decide for itself what rules it will impose on participants. If they choose to limit access to new, safe enhancement biotechnologies, that's perfectly fine. However, it is not at all clear that such restrictions are intrinsically more "moral" or authentic than allowing the use of the new technologies. Perhaps gene doping can be thought of as leveling the playing field, allowing even the genetically less gifted to compete. A playing field on which everyone has optimized their physical abilities would become an arena in which true grit, determination, and character come to the fore. Besides wouldn't it be ironic if the old geezers in the stands were more muscular than the players on the field?

Research so far has concentrated on devising gene therapy fixes for single broken genes, but most common diseases such as heart disease, cancer, and diabetes are caused by a combination of many genes interacting with environmental influences. Researchers are busy poring over the human genome to find the variations that make each individual different from all other persons. They are looking not just for gene variants for traits such as hair and eye color, but also for gene variants associated with disease propensities. Constellations of certain genes are associated with heightened propensities for specific diseases. As difficult as it currently is to fix single gene defects, using small vectors to try to install several genes to reduce a patient's risk of a heart attack would be even more challenging.

One way around the problem might be to insert artificial chromosomes loaded up with the proper suite of genes to counteract those associated with common diseases. Such artificial chromosomes have already been developed by Chromos Molecular Systems of Burnaby, British Columbia, and Athersys in Cleveland, Ohio.[26] Someday, the gene variants that allow some people to eat all the foie gras and pizza they want and still have clean-as-a-whistle coronary arteries will be identified. People who are not so fortunate as to have been born with those heart-protective genes might someday have an artificial chromosome inserted that contains genes optimized for coronary longevity—the foie gras genes, if you will.

FINALLY WINNING THE WAR ON CANCER

Gene therapy targeted at cancer cures is especially promising. After all, cancer is a genetic disease in the sense that it is often caused by the genes that normally trigger the processes that tell damaged cells to kill themselves going awry.

University of California, Berkeley cancer and antiaging researcher Judith Campisi says, "I can make the following statements about ninety percent of cancer cells. Number one, they have some problem with the p53 gene. The cancers may not have a direct mutation in the p53 gene, but somewhere in the p53 pathway, meaning the upstream regulators or the downstream factors there is a problem. It may even be as high 100 percent."[27] P53 is a tumor-suppressor gene that makes a protein that tells a damaged cell to stop dividing. When it mutates or is somehow repressed and thereby becomes ineffective, damaged cells divide uncontrollably and form tumors.

Campisi adds that a very high percentage of human cancers also have problems in the retinoblastoma pathway, another cell-cycle regulatory pathway that controls cell division. The retinoblastoma gene produces proteins that hook into the p53 pathway further downstream, but it's a major regulator of cell growth and a major regulator of senescence. In addition, Campisi notes, somewhere between 80 and 90 percent of all cancers have turned the telomerase gene on, whereas in their normal cellular counterpart it's off.

When normal cells divide, their telomeres—repeated bits of DNA at the ends of each chromosome—shorten. When a cell's telomeres have dwindled away, it signals the cell to cease dividing forever and become senescent, or to commit suicide. Almost all cancer cells somehow reactivate their ability to make telomerase, which enables them to rebuild their telomeres. By maintaining their telomeres, tumors avoid triggering the cellular mechanisms that would normally cause damaged cells to stop dividing or commit suicide. Once telomerase is reactivated, cancer cells are able to grow uncontrollably by dividing indefinitely.

As the cases of Bernis Teaters and Alfredo Gonzalvo illustrated, researchers are now testing gene therapies for restoring healthy genes as a way to kill cancerous cells. In 2003 the company SiBiono GeneTech, based in Shenzhen, China, received approval from Chinese regulatory authorities to commercialize its Gendicine anticancer gene therapy drug.[28] Gendicine, the first gene therapy drug ever to receive such approval, is a viral vaccine in which the p53 tumor suppressor gene has been incorporated into a deactivated cold virus. It was originally approved for the treatment of head and neck squamous cell cancers. A year after an initial clinical trial, 64 percent of patients treated with Gendicine and radiation saw a complete regression of their tumors, a rate three times better than those treated solely with radiation. The chief side effect is transient fever in about a third of the patients.

The drug is generally injected directly into tumors and combined with radiation therapy. Now Westerners such as American businessman Arthur Winiarski, age forty-eight, are going to clinics in China to receive the treatment. Winiarski was diagnosed with squamous cell carcinoma in his sinuses and was told that he had only months to live. He sought treatment with Gendicine at a clinic in Beijing. After undergoing injections of Gendicine and surgery, Winiarski's cancer has disappeared.[29]

Gendicine has proven itself effective at treating a range of other cancers as well. "In clinical trials, Gendicine has been used to successfully treat cancers of the digestive tract (esophageal, gastric, intestine, liver, pancreas, gallbladder, rectum), lung cancer, sarcoma, thyroid-gland cancer, breast cancer, cervical cancer, and ovarian cancer," notes Zhaohui Peng, SiBiono's founder, chairman, and CEO.[30]

Gendicine works by replacing and overexpressing the malfunctioning p53 genes in cancer cells. The new genes then manufacture a protein that orders the cells to stop dividing and to commit suicide. The overexpression of p53 proteins also attracts the attention of immune system cells that then attack the cancer cells. Finally, the restored p53 genes apparently tell VEGF genes, which cause blood vessels to grow toward tumors and supply them with nourishment, to stop working.

In the United States, the Texas-based biotech company Introgen Therapeutics is now moving forward with clinical trials for its own p53 viral vaccine, Advexin. Advexin has shown considerable promise in treating head and neck cancers, lung cancer, and breast cancer. For example, Phase II clinical trials of Advexin for head and neck cancers found that it inhibited or shrank tumors in 41 percent of patients.[31] In a small trial involving twelve women with advanced breast cancer, Advexin was used with chemotherapeutic agents to shrink tumors so that they could be surgically removed. On average, the breast tumors shrank by 80 percent and tumors in lymph nodes decreased by 70 percent.[32] Another small clinical trial involving nineteen lung cancer patients treated with Advexin combined with radiation found that the tumors shrank dramatically in 60 percent of the patients.[33]

The biotech company Geron Corporation in California has developed an antitelomerase vaccine that induces a patient's immune system to seek out and destroy cells expressing the telomerase gene. (Normal cells do not generally express telomerase, so it's a sign that the cell is cancerous.) Thomas Okarma, CEO of Geron, notes that in a Phase I/II trial at Duke University that tested the vaccine for prostate cancer, nine out of ten patients with prostate cancer cells circulating in their blood were able to clear or significantly reduce those circulating cells after receiving six weekly injections of Geron's antitelomerase vaccine. Typically as prostate tumors grow, levels of prostate-specific antigen (PSA) rise in patients' blood. The time it took for PSA circulating in a patient's blood to double rose from three months to more than one hundred months for patients injected with Geron's antitelomerase vaccine. This increased PSA doubling time indicates that the vaccine slows the tumors' growth.

To create the antitelomerase vaccine, patients undergo leukophoresis, in which a machine pulls out white blood cells called dendritic cells from the patient's blood during an hourlong session. In the immune system, dendritic cells capture and present substances (antigens) to immune system cells called T cells. T cells exposed to the antigen then expand in number and become guided missiles specifically targeted to destroy only cells that express that antigen, be they viruses or cancer cells. To make the antitelomerase vaccine, Okarma explains, "We pulse those dendritic cells with the RNA that codes for the protein component of telomerase and that gives them a big jolt of telomerase."[34] The dendritic cells, which now display the telomerase protein, are injected back into a patient as a vaccine. Like all vaccines, the antitelomerase vaccine acts by turning T cells on against cells bearing those proteins. From one blood draw, Geron can produce enough telomerase-exhibiting dendritic cells for twelve to fifteen vaccinations—enough to treat a patient for a year or more. Currently, patients need to be vaccinated once every month or so to keep their T cells active against the cancer cells.

One concern is that telomerase is expressed in the body's stem cells, which continually renew tissues that have a rapid turnover, such as skin, intestinal lining, and blood. So wouldn't antitelomerase T cells attack these vital stem cells as well as tumor cells? Okarma points out that when such stem cells divide they express about one-twentieth of the level of telomerase typically found in tumor cells. So far, evidence indicates that the antitelomerase vaccine does not harm normal stem cells.

As an aside, vaccines developed by exposing dendritic cells outside the body also seem to work on other diseases. For example, a recent clinical trial in Brazil found that a vaccine made by exposing dendritic cells to killed HIV viruses is effective in stopping the progression of AIDS in people already infected with the HIV/AIDS virus.[35]

Since nearly all cancer cells produce telomerase, wouldn't Geron's vaccine be useful against more than just prostate cancer? Okarma agrees: "We're studying it in prostate cancer but there's no reason why it shouldn't work in all patients for all cancers. Other academic people who have used other methods to generate anti-telomerase T cells have shown, as predicted, broad reactivity of those T cells against breast, leukemia, lymphoma, renal cell, and a number of different tumors."[36] As Campisi notes, "If I had a bad cancer, and I was offered a telomerase inhibitor, I would take it."[37]

SHOOTING THE MESSENGER RNA

Ribonucleic acid interference (RNAi) may be one of the most important scientific breakthroughs in decades. In 2002 *Science* magazine hailed

RNAi technology as the number-one scientific breakthrough of the year, and *Technology Review* identified RNAi as one of "Top 10 Emerging Technologies for 2004."[38] RNA interference technology is based on the insight that defective genes don't actually do any damage themselves. They are simply miscopied or garbled information sitting in a recipe database, and as long as they remain unread in the database, they can't cause any harm. Defective genes cause damage when their garbled recipes get cooked up into stuff that can cause harm, generally misshaped proteins. Gene transfer therapy aims to fix the database by substituting the correct recipe before a cook accesses it. By contrast, RNAi technology identifies the garbled recipe after it has been printed out, but destroys the printout before the cook can read it to make poisonous pies.

First, a brief lesson in molecular biology: In multicellular creatures, the information for making proteins is encoded on the double strands of DNA that reside inside the nuclei of our cells, making up our genes. In order to produce proteins, the encoded information must get out of the nuclei and into protein-making machines called ribosomes, which exist in the cytoplasm of cells. To make a long story short, DNA is transcribed inside the nucleus into single-stranded molecules called messenger RNA (mRNA) that carry the recipe for proteins from the nuclei to the ribosomes, which then read and translate that recipe into proteins. So in my analogy, the database is a cell's nucleus, the recipe is the gene, the printout is the mRNA, the cook is the ribosome.

In many disease processes—for instance, cancer—genes can become mutated and thus produce a defective protein recipe, or they can produce too much mRNA and thus too much of a particular protein. As previously discussed, the early idea of gene therapy was that such mutated genes could be replaced with good copies. But as we've seen, this turns out to be very hard to do. So, if you can't replace the gene-gone-bad, why not shoot the messenger—the messenger RNA, that is? This is where RNAi comes in.

It turns out that short sequences of double-stranded RNA (dsRNA) can silence genes by interfering with the messenger RNA produced by any specific gene. The dsRNA consists of two complementary strands of

RNA: One strand has the identical sequence of nucleotides of a specific mRNA strand naturally produced as the targeted gene is transcribed. The complementary strand, called the antisense strand, is a mirror image whose sequence is exactly opposite to that of the specified mRNA molecule.

RNAi is a process of selective gene silencing by destruction of mRNA. It is triggered by dsRNA, in which one strand is identical to the target mRNA sequence. RNAi appears to be an ancient form of defense against viral infection. The genomes of many viruses consist of double-stranded RNA. Such viruses try to hijack a cell's own protein and RNA synthesis machinery to make copies of themselves. The Dicer enzyme protects cells by cutting up any dsRNA strands it encounters into shorter pieces called small interfering RNAs (siRNAs) that are typically less than twenty-one nucleotides in length (DNA and RNA are linear chains of nucleotides). The siRNAs then provoke a response that destroys any viral mRNA that matches the dsRNA trigger.[39] This prevents the virus from taking over the cell and reproducing itself. RNAi is a natural process, intrinsic to every cell of every multicellular organism.

Therapeutic RNAi takes advantage of this intrinsic defensive process by creating synthetic dsRNAs that match mRNAs produced by a cell's own genes. Dicer treats these synthetic dsRNAs as though they were viral sequences and cuts them into double-stranded siRNAs. These siRNAs are then integrated into protein complexes called RNAi-induced silencing complexes (RISCs). RISCs guide the siRNAs to the target mRNA sequence, just as it would have done had the mRNA been produced by a virus. At some point the two siRNA sense and antisense strands unwind and separate. RISCs use the antisense strand of the siRNA to bind to and degrade the corresponding mRNA, resulting in gene silencing.

Stretching my earlier analogy a bit further, RISCs are like paper shredders that destroy the printouts of the garbled recipes. RNAi is astonishingly efficient, because RISCs act as enzymes—proteins that accelerate the rate of biochemical reactions without being themselves changed by the reactions. In this case, RISCs catalyze multiple rounds of RNAi, perhaps hundreds or thousands in each cell.[40]

The therapeutic dsRNAs can be delivered using various vectors. For example, Sirna Therapeutics, headquartered in Boulder, Colorado, has teamed up with Seattle-based Targeted Genetics to develop an RNAi therapy for Huntington's disease. As the Web site of the National Institute of Neurological Disorders and Stroke explains: "Huntington's disease (HD) results from genetically programmed degeneration of brain cells, called neurons, in certain areas of the brain. This degeneration causes uncontrolled movements, loss of intellectual faculties, and emotional disturbance. HD is a familial disease, passed from parent to child through a mutation in the normal gene. Each child of an HD parent has a 50-50 chance of inheriting the HD gene."[41] Disease onset usually begins between ages thirty and fifty. As the disease progresses, walking, speaking, and swallowing abilities deteriorate, leaving the person unable to care for him- or herself. Death typically follows from complications such as choking, infection, or heart failure.

The HD gene is dominant, which means that a person needs to inherit only one copy of it from a parent in order to contract the disease. About 30,000 Americans have Huntington's disease and about 150,000 more are at risk of having inherited the disease from a parent. The disease is caused by defective gene coding of the huntingtin protein. In one section of the gene, a triplet of the DNA bases made up of cytosine, adenine, and guanine (CAG) is repeated many, many times. Normally the CAG triplet is repeated fewer than thirty times. In people with Huntington's disease, the sequence repeats itself dozens of times. The greater the number of repeats, the more likely it is that the person will develop symptoms and the greater the chance those symptoms will appear at a younger age. The disease may occur earlier and more severely in each succeeding affected generation because the number of repeats can increase as the genes are passed down the generations. Probably the most famous victim of Huntington's disease was American folksinger Woody Guthrie, who inherited the disease from his mother. He began experiencing symptoms in his early forties and died of the disease in 1967. There is currently no cure for Huntington's.

Scientists at Sirna plan to counter the production of the abnormal huntingtin protein using RNAi. To do this, they will insert synthetic dsRNA, mimicking a section of the messenger RNA that codes for the abnormal huntingtin protein, into Targeted Genetics' adeno-associated viral vector (AAV). If all goes well, the viral vector will deliver the dsRNA, provoking the process that silences the messenger RNA produced by the defective Huntington's gene. If the abnormal protein is not produced, the people carrying the defective Huntington's disease gene should not become ill.

RNAi technology is not limited to treating inherited diseases. Ken Reed, director of research and technology for the Australian biotech firm Benitec, offers an example of how RNAi might be used to treat cancer. According to Reed, in about 50 percent of all cancers, the p53 gene is inactivated by being repressed by the YB1 DNA-binding protein.[42] This is critical, because p53's role in cells is to tell them to die if they become cancerous. RNAis can be created that will interfere with the mRNA that instructs ribosomes to make YB1 DNA-binding protein. This would free p53 to become effective again at instructing cancer cells to commit suicide. Another biotech company—Intradigm Corporation, based in Maryland—is using RNAi technology to inhibit the VEGF gene, which tumors use to induce the body to grow blood vessels to supply them with nutrients. Cutting off a tumor's blood supply significantly slows down its growth.

RNAi shows a lot of promise for treating infectious diseases as well. After all, RNAi did evolve as a way to defend cells against invading microbes. Researchers are using RNAi technology to develop ways to treat chronic hepatitis B infections by silencing the genes the pathogen uses to produce its outer protein coat.[43] Researchers at the City of Hope Hospital in Los Angeles are planning to use RNAi technology in patients to target HIV/AIDS next year. Researchers at the biotech company Alnylam Pharmaceuticals, based in Cambridge, Massachusetts, and Kulmbach, Germany, took a major step forward in November 2004 when they succeeded in using RNAi to cut cholesterol levels in mice in half. Alnylam's breakthrough was not in curing mice, but in figuring out how to deliver RNAi technology as an intravenous drug. Previously, RNAi therapies had to be delivered directly to the tissues and organs involved. The Alnylam researchers attached siRNAs to a modified cholesterol molecule and then administered the compound intravenously. Cells readily absorbed the siRNA/cholesterol compound and the siRNAs successfully silenced the production of cholesterol.[44]

"This discovery has major potential for the discovery of new medicines," declared Phillip A. Sharp, a 1993 winner of the Nobel Prize in Medicine for his work with RNA and a member of the board of directors of Alnylam. Speaking at a teleconference in November 2004 announcing the results of the study, Sharp added that the successful application of RNAi gene silencing "offers a completely new way of treating disease, as it has the potential to target any gene involved in the cause or pathway of human disease."[45] John Maraganore, CEO of Alnylam, added, "We view our work as a historic step forward in the development of RNAi therapeutics as a potential new class of drugs."[46] Allowing for the usual corporate hype, Sharp and Maraganore may just be right in this case.

In November 2004 Sirna Therapeutics launched its first Phase I clinical trial to treat the "wet" version of a blinding eye disease, age-related macular degeneration (AMD). Wet AMD is characterized by the growth of abnormal retinal blood vessels that leak blood or fluid, causing rapid and severe central vision loss. Sirna researchers are injecting siRNAs designed to inhibit the expression of the VEGF gene, which encourages the growth of blood vessels. Inhibiting the VEGF gene should stop the growth of the blinding blood vessels. If Sirna's RNAi therapy works, it could benefit the more than 1.5 million adults over the age of fifty in the United States who suffer from AMD. The Phase I clinical trial is designed to measure the safety, tolerability, and biological activity of the siRNAs after they are injected into patients' eyes. Nevertheless, we should soon know if RNAi is actually as powerful a treatment as it is being touted to be.

DIAGNOSING THE FUTURE

In 2005 the GeneTests project at the University of Washington estimated that genetic tests had been developed for about eleven hundred health conditions and that eight hundred of them were available to physicians for clinical use.[47] For example, a highly accurate diagnostic test for the Huntington's disease gene has been available since 1987, but only about 20 percent of at-risk Americans have chosen to take the test.[48] Often, people at risk of developing a fatal disease want to know if their time is limited, so that they can make sure not to put off the important things they want to accomplish while still healthy. Since there is no current cure, the majority of at-risk people apparently prefer to live in hope that they won't develop the disease. Many also fear job, insurance, and personal discrimination should they test positive for the deleterious version of the Huntington's gene.[49]

The reaction of at-risk people to the diagnostic test for Huntington's informs us about how people are likely to react as more and more predictive genetic tests become available. Some of us will want access to the most information possible about our health, so we can use that information to manage medical—and especially nonmedical—aspects of our lives. Others, believing that many diagnostic tests simply provide information about conditions that seem inevitable, will choose to forgo such tests. Today we live in a kind of genetic dark ages—patients can learn which diseases they are prone to, but there is often relatively little therapeutic intervention that can improve possible outcomes.[50] However, as the dark ages dissipate and new therapeutic options for various diseases become available, more and more people will seek diagnostic tests so that they can take advantage of the new therapies. For example, if RNAi therapy is able to slow, forestall, or cure Huntington's disease, most people at risk of HD will eagerly choose to take the diagnostic tests.

Unfortunately, the initial myopic response of many bioethicists and physicians has been to discourage patient access to the new genetic diagnostic tests. "Ninety-nine percent of all of our genetic tests tell you something that there is nothing you can do about for yourself," declares George

Annas, pointing in particular to the test for Huntington's disease. According to Annas, the test tells a patient, "You're going to die this horrible death, and there's nothing you can do about it." Thus, he claims, the tests end up creating pathology in people who are healthy but who now think that they are sick.[51]

But while the availability of treatment may be the only relevant consideration for Annas and other bioethicists, patients may have excellent reasons of their own for wanting the information that a biotech diagnostic test might offer. Instead of "creating pathology," the results may well give patients a chance to shape their lives, careers, and reproductive choices appropriately. A man who learns at age twenty that he will contract the dementia of Huntington's in his midforties might decide to lead an entirely different life, skipping the corporate grind for any number of alternatives. Or he might choose not to have children, or to use in vitro fertilization combined with preimplantation genetic diagnosis to select embryos without the lethal gene so as not to risk passing on the disease to his progeny (see chapter 5).

Telling someone that his chances of early-onset heart disease are greater than average may or may not make him feel worse, but it could well energize him to do something about it: get more frequent heart scans, take cholesterol-lowering drugs, exercise more often. The view expressed by Annas and other bioethicists seems to be based on the assumption that everyone is a depressive fatalist.

Gregory Stock, the director of the UCLA Program on Medicine, Technology, and Society, finds the idea that bioethicists would withhold genetic test results from a patient "very offensive." Bioethicists, he says, have "a lack of faith in the individual, essentially."[52] Condescending would-be bioethical mandarins are discounting the well-known and widespread phenomenon of social learning—as new technologies become available, people incorporate and welcome them into their lives. Consider the World Wide Web: in 1992 only four Web sites existed; at the beginning of 2005, Google now searches more than 8 billion Web pages. As diagnostic tests proliferate, more and more people will become increasingly accustomed to handling such genetic information in managing their lives. Genetic information is not some special, mysterious domain of knowledge requiring self-appointed priest figures to protect the uninitiated from its especially corrosive and perilous effects.

In any case, the hand-wringing of today's bioethicists will seem quaint as single genetic diagnostic tests give way to whole-genome scanning. Humanity's extended immune system will eventually develop so that each patient's whole genome, his entire genetic makeup, can be scanned in his doctor's office for less than one thousand dollars. A whole genome scan will alert patients to all of their genetic susceptibilities and let them take preventive action before the disease arises.

Whole-genome scans are the foundation for the development of personalized medicine. Paul Oestreicher of Genaissance Pharmaceuticals declared in 2001 that his company's aim is "to revolutionize health care by customizing treatments based on each person's DNA. Our goal is to eliminate trial and error prescribing."[53] Genaissance identifies genetic variations among individuals, which are organized into haplotypes. Genetic variations affect drug response, so Genaissance can classify patients as to whether or not they will respond to a specific medicine and/or experience side effects.

Oestreicher gave an example of tests for response to the asthma drug albuterol. These tests check patients for genetic variations in a particular receptor on the surface of cells. Scientists at Genaissance found that they could divide patients into four different haplotypes and that two of the haplotypes simply did not respond to albuterol. One day, doctors will be able to test for these haplotypes and avoid prescribing albuterol to those who will not benefit from it. Haplotyping would also be very useful in clinical trials of new drugs, because the drugs could be targeted to patient populations that are likely to benefit. Drug companies know that they have many drugs on their shelves that work for some patients but not for others. Until now they had no way of finding those patients who are likely to benefit. With haplotyping tests, they do.

In December 2004 the FDA approved the first diagnostic test that allows physicians to test patients for genetic differences in metabolization of various drugs for cardiac disease, psychiatric disease, and cancer.[54]

The new AmpliChip Cytochrome P450 Genotyping Test is made by Roche Molecular Systems of Pleasanton, California. This blood test analyzes one of a family of genes called cytochrome P450 genes, which produce liver enzymes that break down certain drugs and other compounds. People are born with different forms of this gene, and some metabolize certain drugs more quickly or more slowly than average, or, in some cases, not at all. This difference in the rate of drug metabolism may explain why some people respond well to antidepressants such as Prozac, whereas others do not. With this test in hand, physicians will be able to tailor dosages for many drugs to fit each patient's genetic profile.

One often-heard concern about the widespread use of predictive genetic tests is that people will risk losing their health and life insurance. Why? Because insurance companies might argue that a propensity toward disease identified by genetic tests is an uninsurable preexisting condition and exclude that disease from coverage. On the other hand, denying insurance companies access to the information provided by genetic tests leads to the problem of adverse selection. Adverse selection occurs when there is an asymmetry of information. In this case, people whose genetic tests indicate that they are at high risk of becoming ill or dying prematurely are likely to load up on gold-plated insurance coverage. Because insurers won't know the results of the tests, the premiums they charge will not cover the cost of taking care of high-risk customers. This means that healthier customers will be charged more to pay for their high-risk counterparts. This situation can set off an adverse selection spiral in which low-risk clients flee the higher premiums, and high-risk clients flock to buy the insurance. As premiums rise to cover the unhealthy clients, fewer and fewer people will be able to afford insurance. One possible solution to this problem would be to let patients buy "genetic test insurance" to insure themselves against any unknowns before they take particular genetic tests.

Genetic testing is not only about disease. Commercial labs now offer cheap, simple, and highly accurate genetic paternity tests through the mail. Simply wipe the inside of a child's cheek with a cotton swab, send it in with $150, and the results are back in a week or so. The simplicity and lower costs boosted the number of paternity tests in the United States from 77,000 in 1988 to nearly 341,000 in 2002.[55] Those seeking paternity tests may have good reason: 28 percent (97,681) of DNA paternity tests in 2002 indicated that the man being tested was not the biological father of the child in question.

With the disdain for patient empowerment that pervades much bioethical thinking, bioethicist Eric Juengst, formerly the director of the Ethical, Legal and Social Implications Program of the US National Human Genome Research Institute, and now at Case Western Reserve University, denounces the new paternity tests, asking, "What is the benefit to the child and society of seeking this information?" "A lot of people have grown up with healthy and stable families in the ignorance of their parental status with no harm done," claims Juengst. He asserts that the information "erodes the . . . unconditional acceptance of our children with all their surprising traits and maybe their genes." He adds that "adoptees can testify that there's a lot more to being a family than just genetic links."[56] Of course, adoptive parents *choose* to rear genetically unrelated children; cuckolded husbands may want the option of knowing what they are doing.

More important, Juengst's view is again typical of how the bioethical community regularly elevates the presumed claims of society over the rights of individuals to choose. All mothers can be sure that the child they bear is theirs. In the future men may enjoy the same certainty for the first time in history as paternity testing at birth of all children becomes routine. Such routine testing may well change the incidence of infidelity among women while also enabling the authorities to identify and seek support from so-called deadbeat dads.

FIGHTING FUTURE PLAGUES—NATURAL AND MAN-MADE

The Genome Age officially began on June 26, 2000, when J. Craig Venter of Celera Genomics and Francis Collins of the National Human Genome

Research Institute announced at the White House that a draft version of the human genome had been completed. The human genome is a listing of all of the 3.1 billion DNA bases that encode the twenty to twenty-five thousand or so genes that are the recipe for making human bodies and minds.

Humanity is already benefiting from the Genome Age. Consider the amazingly fast global response to the severe acute respiratory syndrome (SARS) outbreak in the spring of 2003. The malady infected more than 8,000 people and killed 812 after it broke out in China in November 2002, and many feared that it would become a worldwide pandemic, possibly killing millions. Before the advent of vaccines and antibiotics, infectious diseases largely spread through human populations until genes to stop them arose via natural selection or the diseases became less virulent. That process is slow and painful, to say the least. Just a decade before the SARS outbreak, it would have probably taken scientists years to identify the microbe responsible for SARS.

Prior to 1995 a few genome sequences of viruses such as HIV had been very laboriously and expensively completed. In 1995 the Institute for Genomic Research (TIGR), the private biotech institute headed by Venter in Maryland, jump-started the Genome Age by producing the first sequence of all the genes of a free living organism, the bacterium *Haemophilus influenzae*. This effort took thirteen months and nine hundred thousand dollars.

Over the past ten years, the process of decoding genomes has become routine. In February 2001 six years after the completion of *H. influenzae*'s 1.8 megabase (a megabase equals 1 million DNA bases) sequence, the first rough drafts of the three thousand-megabase human genome were announced in *Science* and *Nature*.

Today, some gene-sequencing machines can read nearly 2 million DNA bases per day, so conceptually sequencing something like *H. influenzae* could take little more than a day rather than thirteen months. Since *H. influenzae*'s genome was sequenced, researchers have sequenced all or parts of the genomes of hundreds of different organisms, from viruses to bacteria to humans. Surprisingly, it turns out that it takes about the same number of genes (twenty to twenty-five thousand) to make a human, a mouse, a pufferfish, a chicken, and a mustard plant.

SARS is caused by an emergent coronavirus, so called because its outer protein coat looks like a miniature crown. Once the virus was isolated from a patient, it was sent to researchers at the Michael Smith Genome Sciences Centre in Vancouver, British Columbia. It took them only six days (April 6–12) to complete the virus's genetic sequence.[57] In Atlanta, the CDC announced its SARS sequence two days later. "Research laboratories can use this information to begin to target antiviral drugs, to form the basis for developing vaccines, and to develop diagnostic tests that can lead to early detection," declared Julie Gerberding, director of the CDC.[58]

At the beginning of April 2003, the CDC released a diagnostic test for the protective antibodies patients develop as their immune systems fight the SARS virus. Just two weeks later, as the SARS genetic sequences were being announced, researchers in Hong Kong, which has been especially hard-hit by SARS, announced on April 18 that they had devised a diagnostic test suitable for use in doctors' offices.[59] The biopharmaceutical giant Roche released a commercial version of a SARS test in July 2003—only four months after the disease organism was first identified.[60] This kind of fantastically rapid progress is possible only because of biotechnological advances made during the eight years prior to the SARS outbreak.

On December 13, 2004, just twenty-one months after international health officials recognized SARS as a new infectious disease, researchers at the US National Institutes of Health (NIH) announced that they were beginning a safety trial of a SARS vaccine. NIH director Elias A. Zerhouni declared, "Our team at NIAID [National Institute of Allergy and Infectious Diseases] has been able to develop this vaccine at an unprecedented pace, using technological discoveries that were not available just a few short years ago."[61]

As the Genome Age matures, and disease processes and human immune responses are better understood, vaccine production will be sped up and become routine. As we have seen, other approaches such as RNAi also offer strategies for devising quick protection against novel disease outbreaks.

THE BEST BIO-DEFENSE IS BIO-OFFENSE

Biotechnologically generated superpathogens, beyond the control of medicine, are a truly horrific thought. And in an age in which democratic countries are facing enemies clearly not averse to shocking new means of warfare, they might be a horrifically realistic one.

In 2004 the US federal government launched the National Science Advisory Board for Biosecurity (NSABB). The NSABB is supposed to oversee dual-use biological research—research that could be used by therapists to cure, but could also be exploited by potential bioterrorists to kill.

The NSABB consists of a committee of twenty-five people, appointed by the US Secretary of Health and Human Services, who are experts in areas such as molecular biology, microbiology, infectious diseases, public health/epidemiology, health physics, pharmaceutical production, veterinary medicine, plant health, food production, bioethics, national security, intelligence, law enforcement, and scientific publishing. The NSABB will advise and exercise oversight over biological experiments that would make human or animal vaccines ineffective; grant resistance to therapeutically useful antibiotics or antiviral agents for humans, animals, or crops; increase the virulence of human, animal, or plant pathogens, or make nonpathogens virulent; make pathogens more easily transmissible or alter their host range; help evade diagnostic or detection methods; or enable weaponization of biological agents or toxins.

The NSABB seems intended to function much like the NIH Recombinant DNA Advisory Committee, which has ruled on the safety of genetics research since the 1970s. All experiments would undergo review by the appropriate institutional review boards first, and if additional questions remain about an experiment's national security implications, the NSABB could conduct a further review and suggest limits on what is done and what is published.

So, with a new federal regulatory agency on the case, are we safe from bioterror now? In reality, this new biosecurity agency will regulate only respectable researchers at universities and corporations in this country, who are not likely to be cooking up some superinfectious version of smallpox to spread through the New York City subway system. Thus, this new federal effort may well be irrelevant to the al Qaeda wannabes and illiberal political fanatics of the future.

Real biosecurity lies elsewhere. As the 2004 National Academy of Sciences (NAS) report *Seeking Security: Pathogens, Open Access and Genome Databases* makes clear, what we ultimately need to defend ourselves against bioterror is a highly sophisticated and robust biotechnology infrastructure.[62] According to the NAS report:

> The problem . . . is not to strike the correct balance between security and openness; that is a false dichotomy—openness has enhanced security in the past and is the best way to ensure security in the future. Instead, the most important task is to be as well prepared as possible to cope with the serious infectious-disease threats that society is sure to face in the coming century, both natural and human-made. The [NAS] committee believes firmly that the policies currently in place for genome data—immediate release and free access—are correct because openness is essential to maintain the progress needed to stay ahead of those who would attempt to cause harm.[63]

The NAS committee further noted:

> Research exploiting the revolution in genomics has an important role to play in increasing our ability to defend against infectious agents of importance to biodefense and in global infectious disease. . . . Extensive sequence comparisons between pathogenic and nonpathogenic organisms, studies of changes in the pattern of gene expression in pathogens and their hosts as they interact, and sequencing of multiple strains of specific pathogens will all contribute to the development of new diagnostics, vaccines, and therapeutics for disease-causing organisms, including those which might be used in a bioterror attack.[64]

In other words, effective biodefense depends not on limiting or restricting access to technology, but on nurturing a robust and open biotechnology.

But before we panic about biowarfare agents, let's think a bit about evolution. If a truly horrific virus or bacteria were easily concocted, it is very likely that Mother Nature would already be generating lots of them. Of course, there are exceptions: The bubonic plague did kill one-third of western Europe's population between 1347 and 1350. So let us assume the worst, that fiendishly clever evildoers could devise some sort of superplague that would kill off some huge fraction of humanity: a plague as deadly as Ebola, more communicable than the common cold, and with a latency period of several weeks to allow it to spread through unwitting populations.

What would it take to counter such a pathogen? A dynamic and extensive diagnostic and biomedical manufacturing system that could deploy multiple levels of defense, including vaccines, new antibiotics, and other novel targeted therapies. To do that, we need to move ahead with innovative biotech.

Fortunately, we are well on our way to developing such a biotechnological infrastructure. The future will see a system in which first responders, perhaps using biolabs on a chip, will be able to decode the genomes of pathogens within hours of an outbreak.[65] Once a genome is decoded, biotechnologists could quickly identify essential metabolic circuits and then design therapeutic molecules to disrupt them, thus preventing the spread of the bioterror agent. For example, Lucy Shapiro of Stanford University School of Medicine scrutinizes the published genomes of pathogens looking for genes and processes that are essential for their survival. Then she and her collaborator, biochemist Stephen J. Benkovic of Pennsylvania State University, design molecules that will disrupt those essential processes and kill the microbes. Shapiro and Benkovic have already identified a set of molecules that kill off a wide variety of bacteria, including *Brucella abortus*, *Francisella tularensis*, and multidrug-resistant *Streptococcus*, *Staphylococcus*, and *Mycobacterium tuberculosis*.[66]

Ever since penicillin was developed in the 1940s, researchers have known that finding such antibiotic molecules is a fruitful approach to killing pathogens. However, identifying such compounds was hit-or-miss because the internal workings of most microbes were "black boxes" into which researchers could not peer. So antibiotic researchers essentially threw thousands of compounds at microbes and waited to see which ones did them in. In the last ten years, the publication of hundreds of different genomes has pried opened those black boxes, allowing researchers to tinker with microbial pathways and find their most vulnerable spots. Examples include antibiotics such as the neuraminidase inhibitors Tamiflu and Relenza, which halt influenza infections if taken shortly after exposure or onset of symptoms. They do this by disrupting a protein that the virus needs to replicate. Similarly, researchers have discovered highly effective compounds, such as adefovir, that block anthrax's deadly edema factor toxin.[67]

A scientific team led by Huw Davies at the University of California, Irvine has developed an ultrafast technique that allows scientists to speed the development of vaccines against infectious diseases such as smallpox, malaria, and tuberculosis. The technique enables researchers to use genome information from pathogens to produce essentially all of their proteins all at once. These proteins can then be rapidly tested to see which ones provoke immune reactions in humans. Such proteins, which themselves cannot cause disease, are good targets to use in formulating vaccines against their associated pathogens. In other words, the technique rapidly identifies microbial proteins that could be quickly transformed into safe vaccines.[68]

In January 2004 researchers at Boston's Dana-Farber Cancer Institute reported that they had discovered a specific antibody that reacted to one of the "spike" proteins of the SARS virus. Perhaps in the future labs will design, test, and manufacture vast quantities of such antibodies to protect people exposed to bioattacks from newly bioengineered pathogens.[69]

It should now be clear that it is vital that biotechnological research and development not be stifled in the name of national security. Unfortunately, it will not be possible to stop future bioterrorists from dreaming up and deploying new bioengineered pathogens. But a robust biotechnology should be able to confine the effects of such attacks to no more than the number of people who are killed by car bombs today. Future bioterrorist attacks will be nightmares for those affected, but they ought not be sufficient to destabilize civilization.

This necessarily brief tour of the horizon of recent biotechnological breakthroughs means that many other exciting areas of progress have been omitted, including such developments as personalized cancer vaccines made using each patient's own cells; the ongoing search for new small molecule cancer medicines such as Gleevec and Avastin; the discovery and development of new wide-spectrum anti-infectives such as lactoferrin; the creation of targeted nanoparticles that deliver diphtheria toxin only to cancer cells; and the production of viruses that attack and kill only cancer cells. The technologies highlighted here may not mature into widely used treatments or they may be superseded by other emerging therapies, but there can be no doubt that biomedical progress will eventually vanquish many diseases, cancer and heart disease among them, sooner rather than later. Advanced therapies such as replacing defective genes through gene transfer therapy and RNA interference—or even better biotechnologies—should become routinely available at your doctor's office over the next couple of decades.

More happily for future generations, the technologies described above are just the leading edge of an accelerating biomedical revolution. Over the next fifty years, the remaining mysteries of how genes, cells, tissues, organs, and pathogens interact will be unraveled. Medicine will increasingly shift from a probabilistic empirical practice—if we do *this*, then we think *that* will occur most of the time—to a mechanically predictive science. In other words, cells, bodies, and minds will no longer be black boxes that are manipulated from outside in a hit-or-miss fashion. Instead, future physicians will be able to predict precisely how any given therapeutic or enhancement intervention will cascade through the various metabolic pathways in a patient's cells and tissues. As the predictive power of biomedicine grows, the current model of time-consuming and costly clinical trials—Phase I for safety; Phase II for dosage setting; and Phase III for efficacy—will fade away. A future researcher will conceive of a new therapy, run it through elaborate in vitro model testing systems and computer simulations, and deliver a newly validated safe and efficacious treatment to patients in days or weeks instead of years.

However, until that happy day arrives, as the cases of Jesse Gelsinger and the French SCID leukemia children show, there will be missteps along the way. The public and policymakers must accept that biomedical pioneering is a risky enterprise. Regulators in the United States and France chose to shut down gene therapy clinical trials after the two cases of leukemia diagnosed in the French SCID kids. In contrast, British researchers decided to proceed with their clinical trials to treat SCID patients.[70] As British gene therapy researcher Adrian Thrasher noted in 2002, "We have sixteen patients who are still alive, and two cases of leukaemia that are responding to therapy. Under conventional protocols we would have expected to have lost four or five of them by now."[71] Gene therapy trials have to be put in the context of all medical research. British commentator Mark Henderson notes, "When Christiaan Barnard pioneered heart transplants, of the first 170 recipients, 146 were dead within three years. The early results of bone marrow transplants were even worse. Both procedures now help thousands of people to live full lives. Yet if these early figures were recorded today for a new technique, pressure groups would clamour for a ban."[72] Savio Woo, past president of the American Society of Gene Therapy and director of the gene therapy institute at Mount Sinai Hospital in New York, agrees: "There won't be any new medical therapies forever in the future if the public expects a new therapy to be completely safe the first time."[73] Such a demand for perfect safety would end up killing far more people than it would save. Just ask Bernis Teaters, Alfredo Gonzalvo, Cynthia Cutshall, and Ashanthi DeSilva if the benefits of biotechnological progress outweigh the risks.

CHAPTER 3

ARE STEM CELLS BABIES?

The Ethics of Making Perfect Transplants

Judson Somerville wants to walk again. So he agreed to work with researchers at the Massachusetts-based biotechnology company Advanced Cell Technology (ACT), who might be able to help him achieve his goal. ACT researchers took skin cells from several patients, including Somerville himself, a forty-four-year-old Texas man who injured his spinal cord in a cycling accident and is now paralyzed from his chest down. Using a technique called somatic cell nuclear transfer (SCNT), researchers removed the nuclei that contain Somerville's genetic material from his skin cells and injected them into enucleated human eggs (eggs that have had their own nuclei removed). A few of those eggs then began dividing in petri dishes.

Although the cloned eggs did not get very far along in this particular experiment, the idea is that someday doctors using such "therapeutic cloning" will be able to transform Somerville's skin cells into nerve cells. These nerve cells would be perfect transplants for him, because they would be genetically identical to Somerville's other cells and thus would not be rejected by his immune system. Therapeutic cloning holds out the promise that transplanted nerve cells will knit up Somerville's broken spinal cord, enabling him to walk again.

The era of regenerative medicine of that sort dawned in November 1998, when Geron Corporation announced that scientists whose work it had funded had isolated the grail of human cell biology: embryonic stem cells. James Thomson, a researcher at the University of Wisconsin–Madison, isolated embryonic stem cells from leftover embryos from fertility clinics.[1] Thomson managed to create five distinct perpetually growing stem-cell colonies. Simultaneously, John Gearhart at Johns Hopkins University in Baltimore isolated primordial germ cells (the precursors to eggs and sperm) from donated aborted fetuses, which also acted like stem cells.[2] These remarkable cells are capable of indefinitely renewing themselves and growing into any of the 220 or so types of cells found in the human body. No other cells can do those things. Researchers believe that therapies derived from embryonic stem cells will one day be used to regenerate damaged or aged tissues and organs, restoring them to full healthy functioning.

"The development of cell lines that may produce almost every tissue of the human body is an unprecedented scientific breakthrough. It is not too unrealistic to say that this research has the potential to revolutionize the practice of medicine and improve the quality and length of life," enthused then NIH director Harold Varmus at a congressional hearing a month after the Geron announcement.[3] "Research on embryonic stem cells holds the promise of not only rebuilding organs but rebuilding them so that they are young again," declared William Haseltine, then CEO of the biotech company Human Genome Sciences, at the third annual conference of the Society of Regenerative Medicine.[4]

Stem cells have this promise because, at the very earliest stages of development, an embryo is an undifferentiated collection of cells rather than distinct blood cells, neurons, skin cells, muscle cells, and so forth. These undifferentiated stem cells can develop into any type of tissue. Such stem cells could one day be used to grow new rejuvenated heart, nerve, pancreatic, or liver cells that would replace tissues damaged by disease. Replacement parts produced using stem cells could extend human life spans by decades, with significantly improved quality.

Currently, biotechnologists investigating stem cells mostly use leftover embryos donated by couples who have undergone infertility treatments. The embryos are grown in laboratory cultures until they reach the blastocyst stage, four to seven days after fertilization. At that point the embryo consists of a barely visible ball, about the size of a dust mote, of about 100 to 150 cells. A blastocyst is a hollow sphere of cells whose outer layer, under ideal conditions, will develop into the placenta while the inner cell mass grows into a fetus. Once the inner cell mass is extracted from the blastocyst, those stem cells can no longer develop into a complete organism.

The first colonies of stem cells derived from the blastocyst are still growing in a culture on a layer of mouse feeder cells that provide the necessary environment to keep them alive and undifferentiated. More recently, new stem cell colonies are being maintained without using mouse feeder cells. Researchers are now trying to learn exactly what molecular signals will guide the development of stem cells into specific tissues. Those signals hold the key to using stem cells to create replacement tissues that could become part of a universal tissue repair kit. Suffer a third-degree burn? Grow some skin cells in a petri dish for a skin graft. Heart attack? Replace the damaged tissue with made-to-order heart cells. Broken back? Fix that right up with a skein of new nerve cells.

Since Geron's 1998 breakthrough, a lot of progress has been made in understanding human embryonic stem cells and learning how to direct their development into useful tissues. For example, researchers are able to stably generate heart cells (cardiomyocytes), liver cells (hepatocytes), osteogenic (bone) cells, blood stem cells (hematopoietic progenitor cells), and a variety of nerve cells from embryonic stem cells. In the meantime, many more colonies of human embryonic stem cells have been derived at laboratories around the world.

STEM CELLS: BORN IN CONTROVERSY

Repairing broken bodies, extending life, and improving individuals' capabilities sound like very good things. But as we have seen, the promises of biomedicine increasingly attract fierce opposition, and potential therapies using human embryonic stem cells are no different. A chorus of influential intellectuals is demanding that new biotechnologies be crushed immediately, and in the United States many legislators are listening. These bioconservatives see in life-saving biomedical research the latest incarnation of human evil. "In the 20th century, we failed to stifle at birth the totalitarian concepts which created Nazism and Communism though we knew all along that both were morally evil—because decent men and women did not speak out in time," wrote the British popular historian Paul Johnson in 1999. "Are we going to make the same mistake with this new infant monster [biotechnology] in our midst, still puny as yet but liable, all too soon, to grow gigantic and overwhelm us?"[5]

Stem cell research obviously promises to significantly advance human health and longevity. And just as obviously, stem cell research is completely entangled with the politics of abortion. It involves the use of embryonic cells and, eventually, the creation of fertilized eggs or cloned embryos that abortion opponents consider full-fledged human beings. To abortion opponents, a blastocyst used to duplicate your heart tissue isn't an extension of your tissue, it's another human being—the equivalent of your identical twin. As Judie Brown, president of the American Life League, told the *Los Angeles Times* about research on embryonic cells, "It doesn't matter if it's done in the womb or a petri dish, it's still killing."[6]

So in an ironic linguistic twist, the pro-death opponents of substantially extending human life spans have found their greatest allies among the pro-life opponents of abortion.

After Geron scientists announced that they'd isolated human embryonic stem cells from donated embryos and aborted fetuses, President Bill Clinton on November 14, 1998, asked the National Bioethics Advisory Commission to look into any ethical issues associated with stem cells. A congressional ban, adopted in 1996, outlawed the use of federal funds for the creation of human embryos for research in which they are "destroyed, discarded or knowingly subjected to risk of injury or death." However, in January 1999 US Department of Health and Human Services general counsel Harriet Rabb artfully concluded that Geron's embryonic stem cells "are not a human embryo within the statutory definition." She based her decision on the fact that the cells "do not have the capacity to develop into a human being, even if transferred to the uterus."[7] Consequently, using them in the course of research does not constitute either harming or the destruction of an embryo. Therefore, she concluded, the NIH could fund research using already derived embryonic stem cells.

This ruling immediately provoked seventy antiabortion members of the US House of Representatives to sign a letter of protest to President Clinton, declaring that the HHS ruling, no matter how artful, violated the congressional ban on funding research on human embryos. As it became clearer that the National Bioethics Advisory Commission was going to recommend that some stem cell research be federally funded, opponents turned up the heat. In July 1999 Sen. Sam Brownback (R-KS) sponsored a Capitol Hill press conference featuring a group of bioethicists, religious activists, and physicians who oppose human embryonic stem cell research. "Human embryos are not mere biological tissues or clusters of cells; they are the tiniest of human beings," asserted the group in a press release.[8]

At the press conference, Edmund Pellegrino, a Roman Catholic physician and bioethicist at Georgetown University's Kennedy Bioethics Center, took aim at even private research efforts such as those sponsored by Geron. He urged that a congressional ban "should be extended permanently to include privately supported as well as federally supported research involving the production and destruction of living human embryos."[9] Although a lot of the current debate in the United States centers on federal funding, the real issue is whether the research should be done at all. The bioconservative opponents of human embryonic stem cell research argue that biotechnologists should instead concentrate on isolating and using stem cells known to exist in adults and in umbilical cords discarded after birth. Such adult and umbilical cord stem cells are the precursor cells that renew the blood and tissues such as skin and the lining of the intestines, and in some cases they are already being used to regenerate tissues.

DO WE NEED TO USE EMBRYONIC STEM CELLS?

So why not avoid the contentious area of embryonic stem cell research entirely and just pursue the use of adult stem cells to treat people? After all, transplants of stem cells from adult bone marrow have been used successfully to treat blood diseases such as leukemia for decades. Leukemia is a cancer in which disease-fighting white blood cells multiply uncontrollably. To treat it, physicians first destroy a patient's own blood-forming bone marrow, then infuse the patient with blood-forming stem cells taken from a closely matched bone marrow donor. And if all goes well, the transplanted stem cells colonize the patient's bones and begin to produce normal blood cells. But sources of other types of stem cells have proven to be more elusive.

Recently, however, researcher Catherine Verfaillie at the University of Minnesota caused considerable excitement among stem cell researchers when she claimed to have isolated adult stem cells from bone marrow that seem to be able to transform themselves not only into blood-forming tissues, but also into a wide variety of other tissues, including heart, brain, lung, liver, and muscle cells.[10] Verfaillie calls her cells multipotent adult progenitor cells (MAPCs). However, Evan Snyder, the director of the Stem Cell and Regeneration Program at the Burnham Institute in La Jolla, California, points out that no other laboratories have been able to replicate Verfaillie's work as yet.[11] Furthermore, new research at Stanford University, the Howard Hughes Medical Institute, and the University of California, San Francisco unfortunately casts doubt on Verfaillie's findings.

The Stanford researchers modified bone marrow stem cells in mice to produce a variety of easily detected proteins. These cells were then injected into other mice. Several months later, the rodents were sacrificed and their tissues closely inspected. The researchers found that the bone marrow stem cells had not turned into heart, liver, or brain cells, but had instead fused with those cell types.[12] The fused cells expressed the proteins from the modified bone marrow cells, and could also be identified because they contained two nuclei, the original one plus one from the bone marrow cells.

Although a disappointment, this finding offers some therapeutic promise, perhaps as a way to revitalize aging cells by smuggling fresh, undamaged genes into them. However, the fused cells do not divide to produce new cells. Thus, they will be of limited use in regenerating and replacing tissues damaged by heart attack, diabetes, or Parkinson's disease.

Even if the existence of MAPCs is confirmed, their therapeutic use is still technically problematic: Can enough of them be created in a reasonable time frame at a reasonable cost? Consider the thought of generating one kilogram of cells (10 trillion cells) starting with a vial containing 10 million cells. Human embryonic stem cells double in less than thirty hours and can produce that number of cells in less than twenty-five days. Because embryonic stem cells can be grown more densely, that number can be produced on about one thousand square meters of tissue culture surface. MAPCs, meanwhile, would take forty days to produce a kilogram of cells and take one hundred thousand square meters of tissue culture surface.[13]

Another nonembryonic option is to use a patient's own adult stem cells to repair a damaged organ—called autologous cell therapies. But this, too, presents some dilemmas. First, appropriate cell types may not exist. (For example, no one has found heart stem cells yet. However, University of California at San Diego researchers led by Kenneth Chien announced in February 2005 that they had isolated "cardiac progenitor cells." Chien was able to grow a few hundred of these cells taken from mice into millions of cardiac muscle cells. After birth only a few hundred cardiac progenitor cells survive and their number decreases with age. Unlike stem cells, which have the capacity for perpetual self-renewal, cardiac progenitor cells undergo a finite number of divisions which limits the number of mature cardiac muscle cells they can produce.)[14] Second, it may be impossible to grow enough cells for effective treatments because adult cells, including adult stem cells, have already aged. Therefore their capacity to grow and expand is already limited. Such aged cells are more likely to experience senescence—that is, simply stop dividing—and to suffer genomic instability, exhibiting deleterious chromosomal rearrangements and duplications.

Another potential source of stem cells is the blood and other tissues of umbilical cords and placentas. In the past these tissues were tossed in the garbage after a child's birth. Increasingly, blood is being syringed out of them and frozen in cord blood banks around the world. In the womb, the umbilical cord and placenta supply a growing fetus with nutrients and oxygen and remove waste. Transplanted human umbilical cord blood stem cells were first used in Paris in 1988 as a substitute for a bone marrow transplant to cure a five-year-old boy of Fanconi anemia.[15] More recently, umbilical cord blood stem cells have been used to cure leukemia in adults.[16] Once the malfunctioning blood-forming bone marrow of a leukemia patient is destroyed by radiation, the patient is infused with umbilical cord blood stem cells. The transplanted cells nest in the patient's bones, where they produce all the healthy components of blood: red cells, white cells, and platelets. Transplanted umbilical cord blood stem cells seem less prone than transplanted adult bone marrow cells to cause immune system problems, possibly because of their immaturity. In any case, since the supply of umbilical cord blood is potentially huge, and as cord blood banks grow, success in tissue-matching to prevent immune rejection (or in this case, graft-versus-host-disease, in which the transplant attacks the recipient's body) becomes increasingly likely for each patient.

Still, to ensure the success of such a transplant, donor umbilical cord stem cells need to be matched as closely as possible with the recipient's immune system. This has led to a once-controversial practice in which parents of a child suffering from leukemia or another genetic blood disease use in vitro fertilization and preimplantation genetic diagnosis to bear a healthy sibling whose cord stem cells will be an immunological match for the ailing elder sibling.

Consider the case of six-year-old Molly Nash, who suffered from the fatal genetic disease Fanconi anemia. The only cure for Molly's disease was a bone marrow transplant from a compatible donor. So her parents, Jack and Lisa Nash, used preimplantation genetic testing to help them bear a sibling who would be a perfect genetic match for her. Using in vitro fertilization, Molly's parents produced thirty embryos that were tested for the disease gene and for transplant compatibility. Only five had the right genetic makeup. The fourth attempted pregnancy resulted in the birth of Adam Nash in August 2000. His umbilical cord blood stem cells were used to replace Molly's defective marrow, and now both children are healthy.[17] Although some more traditional religious groups still condemn in vitro fertilization, 61 percent of Americans in a recent poll by Johns Hopkins Genetics and Public Policy Center approve of using biotech techniques to create "savior siblings"[18] as Molly's parents did. And why not? After all, in vitro fertilization and genetic testing in this case resulted in two loved and healthy children.

A KOREAN STEM CELL MIRACLE?

Researchers around the world are trying to find out if umbilical cord blood stem cells can repair tissues other than blood-forming ones. For example, in late November 2004, South Korean researchers reported that they had used stem cells derived from umbilical cord blood to help a woman with a damaged spinal cord walk again.[19] The researchers injected stem cells from umbilical cord blood into the spine of a thirty-seven-year-old woman named Hwang Mi-soon. Hwang, who has been chair-bound for nearly two decades, took several steps using a walker at a press conference and declared her progress a "miracle." And a miracle it is—the cord blood stem cells were injected directly into her injured spinal cord on October 12; a month and a half later she was able to perambulate somewhat.

"I would be very skeptical of drawing any conclusions from one case with no controls," says Snyder. "The extent of my skepticism is incredibly high based on what is known about the biology of umbilical cord cells."[20] Snyder's skepticism seems amply justified by current scientific results. Heretofore, only lab animals with injured spinal cords have been able to walk again after nerve cells derived from human embryonic stem cells have been installed.[21]

However, Saneron CCEL Therapeutics, a Florida biotech startup, reported in 2003 that infusing human umbilical cord blood stem cells into the injured spines of rats "led to some recovery of function" although "none of the rats walked following treatment."[22] The researchers don't know exactly what the stem cells are doing to promote healing.[23] Earlier studies found that infused umbilical cord blood stem cells could somehow repair stroke damage in the brains of lab rats, though only a few of the cells actually seem to differentiate into cells resembling neurons.[24] One of the South Florida researchers earlier estimated that only one of every million umbilical cord blood stem cells could differentiate into nervelike cells.[25] Another experiment in which a nine-month-old girl was transfused with umbilical cord stem cells in the hope of repairing her brain found, upon autopsy, no nerve cells derived from the cord blood cells.[26]

So why can Hwang walk, however haltingly? Given the failure to find regenerated nerve cells in earlier umbilical cord blood stem cell experiments, what else might be going on? After all, unlike Hwang's case, in which cells were injected into an old injury, all previous umbilical cord blood stem cell experiments involved trying to repair damage very shortly after it occurred. This is because scarring generally blocks access to the damaged area. It's not at all evident how umbilical cord stem cells could get around old scar tissue.

Snyder notes it is possible that the actual injection itself caused changes in Hwang's spinal cord. Or she may have responded to more intense levels of physical therapy received after the transplant. "For all we know, holy water could have been just as effective as the umbilical cord cells," Snyder says.[27]

Engaging in a bit of pure speculation on my part, perhaps what occurred with Hwang is related to a recent finding by researchers at Kansas State University. The Kansas researchers isolated what they believe to be stem cells from the matrix of umbilical cords of pigs and humans, not from umbilical cord blood. The matrix, called Wharton's jelly, cushions blood vessels in the umbilical cord. The researchers have been able to propagate the cells for more than eighty generations and have been able to direct them to differentiate into cells that look a lot like neurons. "Our results indicate that Wharton's jelly cells can be expanded in vitro, maintained in culture, and induced to differentiate into neural cells. We think these cells can serve many therapeutic and biotechnological roles in the future," says Kathy Mitchell, the lead researcher on the project.[28] Perhaps the umbilical cord blood stem cells used by the Korean researchers also contained the apparently more versatile stem cells derived from Wharton's jelly?

On September 11, 2001, the National Research Council of the National Academy of Sciences issued a report, *Stem Cells and the Future of Regenerative Medicine*, that argued that somatic cell nuclear transfer research—that is, therapeutic cloning to create immunologically compatible embryonic stem cells—should be "actively pursued."[29] As the report concluded, stem cell–based therapies could alleviate much of the suffering of the 58 million Americans who will be struck in their lifetimes with cardiovascular diseases, the 30 million who will come down with autoimmune diseases, the 16 million who endure diabetes, the 5.5 million who will lose their minds to Alzheimer's, and on and on. Just as the medical revolution ushered in by vaccines and antibiotics vanquished many of the diseases that killed young people in the past century, stem cell therapies may conquer many of the diseases of old age in the twenty-first century.

Judson Somerville, our paralyzed Texan who tried somatic cell nuclear transfer, hopes that therapeutic cloning will someday produce perfect transplants to repair his damaged spinal cord. As we know from Dolly the cloned sheep, factors in egg cytoplasm can reset an adult cell nucleus, giving it the ability to grow into an embryo as a source for stem cells. Using cloning technology, doctors might one day take the nucleus of one of your skin cells, put it in a human egg from which the nucleus has been removed, and allow that cell to divide to the blastocyst stage. They would then take out the stem cells from its inner cell mass and dope them with the appropriate hormones and proteins to turn the stem cells into, say, heart tissue, which could then be used to repair your ailing heart. Using your own cells in this way would mean that your immune system wouldn't reject the newly engrafted tissues, since the tissues would be a perfect match.

As elegant as the above procedure sounds, there are still a number of challenges that must be overcome before embryonic stem cells can be used to cure patients. The first problem with cloned embryonic stem cells is that they are still hard to produce. South Korean scientists announced in February 2004 that they had taken the first step toward creating genetically matched cells and tissues for transplant by growing stem cells from a cloned human blastocyst.[30] The team, led by Hwang Woo-suk of Seoul National University, created embryonic stem cells by means of somatic cell nuclear transfer—which, you will remember, is when the genes from donated human eggs are removed and then adult cells with all their genes are merged with the enucleated eggs.

Unlike the earlier experiment in which Somerville participated, the Korean scientists were able to generate cloned human embryonic stem cells. The Korean researchers first collected 242 eggs from sixteen female volunteers. The scientists were able to coax thirty cloned embryos to develop to the blastocyst stage, consisting of about one hundred cells. (These blastocysts are clones of the adults that donated the genetic material.) They then removed the inner cell mass from twenty of the embryos and were able to establish only one colony of self-perpetuating embryonic stem cells. These cells from the inner cell mass can differentiate into all the diverse types of tissues that form the human body.

CHIMERICAL CURES?

But are such cloned embryonic stem cells truly immunologically compatible? In a proof-of-concept experiment, in June 2002, researchers at the Boston-based ACT reported the results of an experiment in which they created cloned cow embryos by taking genetic material from one cow and combining it with an egg donated from another cow. The idea was to see if the mitochondrial genes derived from the enucleated eggs would provoke immune rejection when cloned cells and tissues were transplanted back into the animal that donated the adult nuclei. Because they cannot yet derive and maintain cow stem cells, the researchers allowed the embryos to grow into small fetuses and then harvested various tissues such as heart and kidney cells. They then transplanted the cells into the cow that had donated the genetic material. The very good news is that the cells were not rejected by the host cow's immune system: The newly installed cloned cells thrived.[31]

However, in light of the number of human eggs it took to produce just one cloned stem cell colony, it is evident that producing cloned embryos and stem cells is highly inefficient using current techniques. Fortunately, there may be various ways around this impasse. To get around the egg shortage, ACT has tried fusing nuclei taken from adult human cells with cow eggs whose nuclei has been removed. Although the work was never submitted to a scientific journal for publication, ACT researchers claim that the fused entities did begin dividing, and that one blastocyst grew to around four hundred cells before they destroyed it.[32]

In 2002 researchers from the Maria Infertility Hospital Medical Institute in Seoul, South Korea, claimed that they had created stem cells using cow eggs. "We succeeded in cloning embryos with human genes by transplanting a human cell nucleus into a cow's egg without a nucleus," said Park Se-phil, who headed up the effort. "The artificially cultivated embryos have ninety-nine percent of human genetic characteristics."[33] This claim should be treated with caution because since 2002 there have been no further announcements from Park.

In 2003 Hui Zhen Sheng of Shanghai Second Medical University published work in a little-known journal claiming that his team had succeeded in fusing skin cells taken from the foreskins of two five-year-old boys and two men, and facial tissue from a sixty-year-old woman with enucleated rabbit eggs. The only genetic material contributed by the rabbit eggs were mitochondrial genes. The Chinese researchers claimed that out of four hundred of these fused entities, one hundred developed to the blastocyst stage, from which the researchers recovered embryonic stem cells.[34] However, other laboratories have been unable to reproduce the Chinese team's work. Many leading stem cell researchers, such as Roger Pederson at Cambridge University, now doubt that the Shanghai laboratory actually succeeded in producing rabbit/human embryonic stem cells.[35]

The researchers were hoping that easily obtainable animal eggs would be useful for producing an endless supply of stem cells. Furthermore, producing these stem cells involved mixtures of animal and human cellular components that might avoid the ethical objections that some have with regard to creating cloned fully human embryos. But this was not the case. These experiments provoked considerable hand-wringing among some conservative antibiotech intellectuals. In a December 2001 opinion piece in the *Wall Street Journal*, William Kristol, editor of the *Weekly Standard*, and Eric Cohen, now editor of the neoconservative technology policy journal the *New Atlantis*, were horrified that "in trying to make human beings live indefinitely, our scientists have begun mixing our genes with those of cows, pigs, and jellyfish."[36]

J. Bottum, the literary editor for the *Weekly Standard*, finds it repugnant that the Japanese government is permitting "human cells to be implanted into fertilized animal eggs for research purposes." Such sinister experiments, says Bottum, could lead to the creation of "a new race of subhuman creatures," possibly even "pig-boys and monkey-girls."[37] Kristol, Cohen, and Bottum are misrepresenting experiments that aim at producing therapeutic benefits for living people, not the creation of animalized humans or humanized animals. Cow, pig, and rabbit eggs are far more available than are human eggs. If transplantable tissues could be created this way, millions of people might be helped.

There are other examples of researchers mixing species genes in search of medical benefits. For example, biotechnologists are adding a few human genes to the genomes of animals such as pigs and cows in order to create therapeutic proteins. This could allow, for instance, cows to produce human insulin in their milk. Currently we produce human insulin by adding human insulin genes to bacteria. Before that technique was developed, a combination of cow and pig insulin was used to treat diabetes. That insulin was rendered from pigs and cows at slaughterhouses. So which is better, producing human insulin in the milk of a herd of contented Holsteins grazing the Vermont countryside, or from their pancreases at slaughterhouses? And why aren't these bioconservative intellectuals outraged that researchers have "humanized" lowly bacteria by adding human genes to them?

Another attempt to breach the animal/human barrier is involved in research that adds a few genes controlling human immunological responses to tissues derived from pigs. The hope is that this will make certain pig organs, for example, hearts and livers, more acceptable to human immune systems.[38] Bottum contemptuously dismisses this medical research as producing "living meat lockers for transplantable organs and tissues."[39] Evidently, in Bottum's twisted sense of morality, if animals are living meat lockers for steaks and pork chops that's all right, but if animals can be biotechnologically tweaked by the addition of a few human genes so as to provide lifesaving transplantable hearts and livers, well, humanity has crossed the line to eternal damnation.

Current biotech research is *not* aimed at creating half-human/half-animal slaves. But might it be abused this way in the future? Just how biologically credible are the scary scenarios being sketched out by Kristol and his cohorts? Old-fashioned low-tech crossbreeding (such as the kind that makes mules from the mating of horses and donkeys) of humans with animals simply won't work to meet any imagined demand for subhuman slaves. The delicate orchestration of embryonic development it takes to produce a live creature would be disrupted very early on in the process, because the proteins and genetic instructions between a pig and human would be incompatible. So, no pig-boys, then.

But aren't primates such as chimpanzees fairly close in evolutionary terms to human beings? After all, J. Michael Bedford reported in the May 28, 1981, issue of *Nature* that human sperm can penetrate the outer membranes of healthy gibbon eggs.[40] So what about producing monkey-girls through crossbreeding? Clearly, any attempt to crossbreed primates and humans would be odious and should be outlawed, if it's not already. But again, this has nothing to do with medical biotechnology.

What about a more high-tech combining of embryonic cells from humans and animals to produce what bioresearchers call *chimeras* (after the Greek mythological beast that was part lion, goat, and serpent)? This has been done between sheep and goats—though the results are always sterile. How about a chimerical pig-human, or "piman"? Most researchers think that this is biologically impossible because the developmental programs of human cells and pig cells are so different that any attempted combination would simply fail to survive.

Then what about cloning humans using an enucleated cow egg to jump-start the process? As we've seen, some researchers hope to produce human-compatible stem cells using this technique. However, no one knows if the process could result in a live baby or not, nor what effect it would have on the health of any such baby. Since it is not safe, it would be unethical to try to use this technique to produce a baby.

STEM CELL FACTORIES

These bioconservative intellectuals have confused *being human* with merely *having human DNA*. They are treating human DNA as though it were somehow sacred. But DNA is merely the chemical on which the digital code for making proteins is inscribed. Inserting a human gene in a pig or a petunia is not an act of sacrilege. Human DNA in a pig or petunia will make a protein, not a human being. Human beings really are more than the recipe it takes to make them. Kristol and his cohorts are merely conjuring up these dystopic visions of Doctor Moreauesque half-human, half-animal creatures in an effort to frighten scientifically uninformed policy makers and voters into outlawing biotech research they oppose.

In any case, using animal eggs as a way to jump-start the production of human-compatible stem cells has apparently failed to work so far. But there might just be an endless supply of human eggs available, although we've been pointed on the way to this result through animal experiments. Researchers led by Hans Sholer at the University of Pennsylvania's School of Veterinary Medicine reported in May 2003 that embryonic stem cells taken from mice can be transformed into mouse eggs.[41] If it works for mouse cells, many researchers believe that it will work for human cells, too, allowing already existing colonies of human embryonic stem cells to be transformed into an unlimited supply of human eggs. Those eggs, in turn, contain the factors needed to jump-start the process that produces malleable embryonic stem cells that can be transformed into any type of cell in the body. Thus, we may have the possibility of a self-sustaining cycle of stem cell production from therapeutic cloning that does not depend on donor eggs from women.

This would eliminate the major technical bottleneck for cloning human embryonic stem cells: the scarcity of the human eggs needed to jump-start the process. Heretofore, human eggs have been harvested from women undergoing hormonal treatments that cause them to superovulate. These treatments are generally unpleasant and may be dangerous to the women's health. Obtaining millions of eggs for stem cell therapies in this way would be neither moral nor practical. Bioconservative opponents of human stem cell research paint lurid scenarios of women confined to human egg farms, forced to produce eggs for avaricious doctors and corporations. Embryonic stem cell opponent Wesley Smith, from the antievolution Discovery Institute in Seattle, calculates that "[o]btaining [800 million] human eggs for [curing diabetes] would involve stimulating the ovaries to hyper-ovulate, which generally produces 7–10 eggs. Assuming a liberal 10 eggs harvested from each procedure, 80 million women of childbearing age would be needed as donors."[42] Feminist Judy Norsigian, the executive director of the Boston Women's Health Book Collective, joins with conservative opponents, citing fears that stem cell therapies would lead to a "massive expansion in the use of women as paid 'egg producers.'"[43] But thanks to the new mouse stem cell work, donated eggs may not be necessary.

The discovery that embryonic stem cells can be transformed into eggs also has implications for human reproduction. It could even make human reproductive cloning irrelevant. How? Proponents of reproductive cloning have always believed that even if it can be made safe, it would nevertheless be a little-used niche treatment for infertility. Now researchers agree that it should be possible to transform embryonic stem cells not only into eggs, but into sperm as well. One possible future scenario would have an infertile couple using eggs derived from previously existing stem cell lines to produce their own embryonic stem cells. These new stem cells with the genetic material derived from the man and the woman could be transformed into eggs and sperm that could then be combined to form an embryo. The new embryo—containing genes from both the man and woman, just like a conventionally produced embryo—could then be implanted in the woman's womb. Gay couples could also use the same technique to produce children sharing both their genes. Of course, this should only be done once the technique has been proven to work safely.

BANKING ON STEM CELLS

Thomas Okarma believes that therapeutic cloning isn't the way to go for producing better organ transplants, anyway. It could work, but would be ultimately too expensive and time consuming to be an efficient medical technique. "It's like bone marrow transplantation and there's no company that's built around bone marrow transplantation," Okarma says.[44] Instead of creating stem cells for each individual patient, Geron proposes to create master cell banks of already derived embryonic stem cells that can be used to treat thousands of patients. The advantage is that such stem cells and the cells derived from them can be standardized and production can be scaled up to produce therapeutic quantities for thousands of patients. But won't transplanted cells and tissues derived from various embryonic stem cell lines be rejected by the immune systems of patients just like regular transplants are?

Not necessarily, says Okarma. Why? Because recent research at Geron has established that purified embryonic stem cells are themselves somewhat immunosuppressive. This property of stem cells is not surprising, since blastocysts implant in uterine walls and are not rejected. Okarma cites an experiment called mixed lymphocyte reaction in which immune cells from one person are mixed with immune cells from another person. Typically, this is a test to see how likely a possible transplant patient is to reject a donated organ—if lymphocytes start to proliferate, this indicates a strong likelihood of immune rejection. "If you dump embryonic stem cells in that cell culture, it prevents your cells from reacting to mine," says Okarma. "Some of those properties are retained by some of the early differentiated descendants of the undifferentiated embryonic stem cells. The point I'm making is that the cells themselves are at a low immunogenic potential." He concludes, "What's been bandied about in the press as the reason why embryonic stem cells will never make it into the clinic is hogwash. It's not going to be anything like the problem the pundits have predicted."[45]

Many supporters and critics of embryonic stem cell research have claimed that clinical applications are decades away. However, Geron plans to file an investigational new drug application with the FDA requesting permission to begin clinical trials using glial cells derived from embryonic stem cells to repair damaged spinal cords in 2005 or early 2006. (Glial cells are the cells that surround and support nerve cells.) Geron's research shows that transplanted glial cells derived from human embryonic stem cells repair the crushed spinal cords of rats by encouraging the restoration of protective myelin sheathing to damaged nerves. (Myelin acts like insulation that keeps nerves from short-circuiting.)

After treatment, paralyzed rats regained sensation and most of their ability to walk using their hind legs. Geron's plan is to transplant the glial cells into patients who have recently had their spinal cords crushed. The first trial would focus on patients whose lower extremities are paralyzed and who have lost bladder control. If all goes well with those trials, the second part of the trial would then treat patients with spinal injuries in their necks that force them on respirators. If such patients can regain their ability to breathe on their own, that would be a massive proof of the success of the treatment. Okarma notes that Geron's application to the FDA will be "five years before our critics ever said embryonic stem cells would hit human trials."[46]

Unfortunately, Judson Somerville would not qualify for the Geron trials, since the embryonic glial cells must be transplanted before significant scar tissue has developed—not much more than two weeks after the injury has occurred. Since the brain and spinal cord are "immune privileged," that is, they exclude most of the body's immune cells, patients receiving glial cell transplants will likely not need to take immune suppressive drugs.

That will be fine for spinal damage, but what about future stem cell transplants for other organs and tissues? Since Geron's embryonic cells are derived from only a few colonies, they won't be exact matches for most patients. So won't patients need to take immune suppressive drugs to prevent their bodies from rejecting them? Not necessarily. To avoid immune-rejection problems, Geron is exploring the idea of chimerizing transplant patients. Geron is developing several stable lines of stem cells—if all goes according to plan, each line will produce on an industrial scale a whole range of useful cells, including heart cells, liver cells, pancreatic islet cells, nerve cells, and so forth, along with hematopoietic stem cells (the blood and immune cell–forming progenitors). The key idea is that before transplanting cells derived from embryonic stem cells, Geron would first give a patient a low dose of immune suppressant and then infuse her with hematopoietic stem cells derived from one of its stem cell colonies. These hematopoietic stem cells would take up residence in the patient's bone marrow, so the patient would produce a mixture of native and transplanted immune cells. Thus, the patient becomes a chimera; that is, her body now consists of two tissues of different genetic compositions.

Once installed, the new hematopoietic cells should trick the patient's body into accepting other types of embryonic stem cells taken from the same line from which the hematopoietic stem cells were derived. In other words, the patient's body now recognizes the new stem cells as part of its self. If this works, stem cell transplants would not require immunosuppression to avoid rejection.

There is good reason to believe that it might work. A version of such tolerization is already being tested in conventional transplant therapies. For example, the FDA approved a clinical trial at the University of Louisville in Kentucky in which transplant patients receive doses of donor bone marrow after undergoing kidney and heart transplants.[47] The procedure has already worked in mice, pigs, and primates.

Another kind of chimera may also help produce compatible tissues and cells for transplantation into people. Human organs and tissues can be grown in animals.

Esmail Zanjani, a professor of medicine at the University of Nevada at Reno, and his colleagues have created animal/human chimeras by injecting human adult stem cells and embryonic stem cells into fetal sheep about halfway through their gestation. The stem cells are transformed into various types of adult human cells and begin to populate the sheep's organs and tissues.[48] The research was originally aimed at finding out if it would be possible to transplant stem cells into developing human fetuses to correct defects in utero. *Chimera* in this case means a creature made by injecting embryonic or bone marrow stem cells from one species into the developing embryo or fetus of the member of another species. Zanjani found that about halfway through fetal development, sheep fetuses' immune systems couldn't recognize foreign cells. Thus, the fetuses happily accept human cells and incorporate them into various organs, where they develop and multiply. Guided by the sheep's own growth signals, the human stem cells develop into liver, heart, skin, or other types of cells. Zanjani reports that in some cases as much as 80 percent of the cells in a sheep's liver are human.[49]

Zanjani's results indicate the promise for a potentially powerful therapy for humans with damaged organs. A physician could take bone marrow stem cells from a patient whose liver is failing and inject them in a sheep or pig fetus. After the lamb or piglet is born, the physician could then harvest human liver cells that would be perfectly compatible with the patient's immune system (after all, they are descendants of the patient's own cells) and install them to repair the damaged liver. This technique might also repair hearts after heart attacks or cure diabetes by restoring islet cells in pancreases. Further down the line, it might be possible to inject human stem cells at just the right stage of development in the sheep or pig fetus so that essentially whole human organs are created.

Researchers will have to evaluate the dangers of possibly transmitting animal viruses when cells and organs from animal/human chimeras are transplanted. Several hundred patients from around the world have already undergone either cellular, whole organ, or extracorporeal xenotransplantation (transplantation between species) without significant transmissions of either infectious viruses or endogenous retroviruses

(viruses incorporated into the genomes of animals), according to a 2003 report to the European Commission on Xenotransplantation.[50]

For example, in 1997 seventeen-year-old Robert Pennington went into a coma from liver failure and needed a transplant. To gain time to obtain a usable organ, his physicians cleansed his blood by passing it through a genetically modified pig's liver outside his body. Pennington recovered and received a transplant. There is no sign that he picked up any diseases from the pig's liver.[51] Clearly, if it turns out that growing human tissues and organs in human/animal chimeras is really feasible, we could avoid the contentious ethical debate over creating transplantable stem cells using human embryos. Of course, animal rights activists will object, but surely if one can kill a sheep for chops or a pig for a ham, such animals can be sacrificed to obtain organs that could keep a human being healthy and alive.

EVERY GENE IS SACRED

One would think, then, that only radical animal rights ideologues could object to research into chimera organs. Alas, the frontiers of biotech research are never so peaceful. Senator Brownback, a fierce opponent of cloning and embryonic stem cell research, wants to introduce legislation that would ban the mixing of human and animal genetic material.[52] Back in 1997 antibiotech activists Jeremy Rifkin from the Foundation on Economic Trends and Stuart Newman, a professor of cell biology at New York Medical College in Valhalla, New York, and a member of the board of the Council for Responsible Genetics, filed a patent application for "chimeric embryos and animals containing human cells."

The patent the men seek claims the rights to every possible mammal-human combination, both in embryonic form and as viable animals. They want the patent so that they can block any such chimera research. If they get it, they intend to refuse to license anyone to conduct any such research, out of their larger ideological objection to any corporate development of biotechnology.[53]

This attitude is creeping into popular journalism. In a semihysterical article in the January/February 2004 issue of *Mother Jones*, left-leaning journalist Mark Dowie reprises the storyline earlier peddled by the conservative *Weekly Standard* about the gruesome possibility of these new biotech methods being used to create half human/half primate slaves. "Could one animal cell make a being suitable for ownership, forced labor, and medical experimentation, just as 'one drop' of black blood once did?" he breathlessly asks.[54] That is over-the-top rhetoric. Just as a gene for human insulin in an *E. coli* bacterium does not make it human, an animal cell or animal gene for insulin in a human being would not make that human an animal.

"Using human embryonic cells and stem cells" for chimera research, Dowie asserts, is "likely to raise the ire of anyone who believes life begins at conception or that human cells are more sacred than those of other creatures."[55] That's not at all clear. First, as noted before, injecting adult stem cells into animal embryos to produce human tissues and organs should in fact make right-to-lifers happy, since it eliminates the need to use human stem cells from embryos for the same purpose. Second, who in the world believes that "human cells are more sacred than those of other creatures?" Human beings are more sacred than other creatures, but I neither worship my skin cells nor venerate my gall bladder.

So far the US Patent Office has refused to issue Rifkin and Newman their desired patent on the grounds that "[s]ince applicant's claimed invention embraces a human being it is not considered to be patentable subject matter."[56] It is far from clear what "embraces a human being" means. A human gene is not a human being; a human chromosome is not a human being; a human cell is not a human being; a human organ is not a human being (except, perhaps, for a human brain). Rifkin and Newman are challenging the Patent Office's refusal to grant them the patent, pointing out that several patents involving human genes and cells have already been issued.

Please keep in mind that Zanjani's chimeric sheep don't look like sheep-men or men-sheep; they look like sheep. They bleat just like sheep and they chew their cud just like sheep. They are just sheep—but sheep that could offer hope to thousands who need organ transplants. But if Rifkin and Newman had already been awarded the patent they seek, it is likely that Zanjani's hopeful research could not have gone forward. One has to wonder what is more immoral—working to provide transplants for sick people, or blocking the development of such transplants?

Besides their role in regenerative medicine, human embryonic stem cells and tissues derived from them could replace animal testing of pharmaceuticals. Researchers could see directly what effects, both healing and deleterious, pharmaceuticals have on human cells. Embryonic stem cell lines derived from embryos with genetic defects would also illuminate how these diseases develop at the cellular level. Yury Verlinsky, the head of the Reproductive Genetics Institute in Chicago, is a pioneer of preimplantation genetic diagnosis. His patients who choose to implant only embryos without genetic diseases have donated hundreds of embryos with genetic defects. Verlinsky has now developed eighteen new lines of disease-carrying embryonic stem cells and is offering them to researchers who want to study their potential for treating inherited diseases.[57]

Opponents of cloning research correctly point out that the techniques devised by the Korean researchers can be used to try to create cloned babies as well. After all, fertility doctors already implant conventionally produced blastocysts into the wombs of infertile women so they can bear children. Given the myriad health problems that cloned animals suffer,[58] it would be unethical to attempt to produce a cloned human baby now. But banning therapeutic cloning just because someone might use its techniques to try to clone a baby is throwing the baby, cloned or not, out with the bath water. Lawrence Goldstein, a researcher from the Howard Hughes Medical Institute, complains that bioethicists love to talk about banning the precursors to any activity they think might be harmful. He thinks this is silly. "I could use a hammer to wage war on my fellow citizens, but we don't ban hammers," Goldstein notes. "A wise society penalizes the acts it wants to prohibit. If we went around banning the precursors to all types of undesirable activities, we would never get anything done."[59] Goldstein sensibly concludes that we might ban reproductive cloning while permitting cloning to produce therapeutic stem cells.

THE POLITICAL WAR ON STEM CELL RESEARCH

Much progress is being made in understanding how stem cells from various sources—bone marrow, umbilical cord, and embryonic—might be developed into useful treatments. Yet the most promising area of research has already been limited in the United States. In his first nationally televised speech on August 9, 2001, President Bush addressed whether the federal government should fund embryonic stem cell research. The president acknowledged that "scientists believe further research using stem cells offers great promise that could help improve the lives of those who suffer from many terrible diseases—from juvenile diabetes to Alzheimer's, from Parkinson's to spinal cord injuries. And while scientists admit they are not yet certain, they believe stem cells derived from embryos have unique potential." Bush also told the nation that he believes "human life is a sacred gift from our Creator" and that he worried "about a culture that devalues life."[60] Thus, because embryonic stem cells are derived from human embryos, he decided to limit federal funding on human embryonic stem cell research to only those cell lines that had been derived before the date of his speech. The president estimated that sixty such stem cell lines would be usable. However, as of December 2004, only twenty-two stem cell lines have been approved by the NIH for funding.

The US House of Representatives has twice voted for a ban on all human cloning research, including therapeutic cloning to produce transplantable cells and tissues. The Human Cloning Prohibition Act would criminalize therapeutic cloning research by fining scientists and physicians up to $1 million and throwing them into prison for up to ten years. The majority of the President's Council on Bioethics recommended a four-year moratorium on therapeutic cloning research. The only American researcher to work with the South Korean researchers to produce the first stem cell colony by means of therapeutic cloning is Jose Cibelli, a professor at Michigan State University. Cibelli lives in a state that has criminalized human cloning research with fines of up to $10 million and jail time of up to ten years. At least one American stem cell scientist, Roger Pederson, left the United States to continue his work in Britain.[61]

The Bush administration pushed for three years to get the General Assembly of the United Nations to adopt a resolution calling for a global treaty to ban all human cloning research, both reproductive and therapeutic. Fortunately this proposal was defeated in November 2004. The result of this push was a nonbinding General Assembly resolution in February 2005 that asks member states to prohibit any form of human cloning that is incompatible with human dignity.

Considering the attempts by the Bush administration and Congress to limit various aspects of stem cell research and therapeutic cloning, one might think the American public was against such research. However, a Harris Poll conducted in August 2004 found that 73 percent of Americans favor using embryos left over from infertility treatments for stem cell research.[62] Opponents argue that supporters of embryonic stem cell research have bamboozled the public. "A lot of people are confused about this issue," says Kris Mineau, president of the Massachusetts Family Institute. "But the more informed they become, the more resistant they are to the idea of the wanton destruction of embryos."[63]

The Harris Poll results show just the opposite. In 2001, 68 percent of people polled had heard anything about the debate over stem cells. By 2004, 83 percent had. In 2001, 61 percent favored human embryonic stem cell research. As we've seen, that rose to 73 percent in 2004. It appears that the more Americans know about the research, the more they support it. Of particular interest to conservative opponents of stem cell research should be the fact that 60 percent of self-identified Republicans want the research to proceed.

The Harris Poll results are bolstered by other polls. In August 2004 the National Annenberg Election Survey found that 64 percent of respondents favored federal funding of research on diseases such as Alzheimer's using stem cells taken from human embryos.[64] Another poll in October 2004 by Virginia Commonwealth University asked respondents if they favored research on stem cells taken from discarded embryos. Fifty-three percent said yes.[65] Opposition to human embryonic stem cell research by bioconservatives in Congress and the White House will turn out to be a losing political proposition.

While the United States has been embroiled in stem cell politics, other nations are moving ahead with the research. In August Britain granted its first license to a University of Newcastle laboratory to create cloned human embryos for the purpose of extracting stem cells.[66] Other countries proceeding with human stem cell research include Japan, Israel, South Korea, China, and Australia.

And in November 2004, 59 percent of California voters passed an initiative authorizing the state to raise and spend $3 billion for stem cell research over the next ten years. Under Bush administration limits, the federal government spent only $24.8 million on human embryonic stem cell research in 2003. The British government has allocated $30 million for stem cell studies in 2004. So with $300 million per year to dispense, the new California Institute for Regenerative Medicine is now the largest government supporter of stem cell research in the world. But California is not alone in the United States. New Jersey has begun a stem cell institute at Rutgers University with $6.5 million in seed funding and a promise of another $43.5 million over the next four years. Wisconsin is considering a plan to spend $750 million dollars over the next ten years on human embryonic stem cell research.[67] To get around federal government restrictions, leading universities such as Harvard and Stanford have established privately funded stem cell research institutes.[68]

THE RELIGIOUS WAR OVER STEM CELL RESEARCH

As noted earlier, opponents want to completely halt or at least limit human embryonic stem cell research largely because they regard it as immoral. It is immoral because they believe that embryos are full members of the human family. For Roman Catholics, human embryos are in fact full human beings and merit all the protections accorded to a healthy thirty-year-old woman. "The human being is to be respected and treated as a person from the moment of conception; and therefore from that same moment his rights as a person must be recognized, among which in the first place is the inviolable right of every innocent human being to life," is the definitive dogma of the Roman Catholic Church on the matter.[69]

Many conservative Protestant churches agree with this view. For example, the Ethics and Religious Liberty Commission (ERLC) of the Southern Baptist Convention issued a statement on October 26, 2004, declaring that a human embryo "is a human life, no matter his or her age, manner of conception (natural conception, in vitro fertilization or cloning) or location (uterus, test tube or Petri dish)." The statement added, "It is incumbent on a just society to protect the lives of these little ones and to search for alternative sources of stem cells."[70] However, other Christian denominations have reached the conclusion that using embryos to produce stem cells is moral. In the United States the Presbyterian Church and the Episcopal Church have both passed resolutions supporting human embryonic stem cell research.[71]

Non-Christian religious traditions have other views about the moral status of human embryos and stem cell research. In Jewish moral thought, during the first forty days of gestation, the fetus, according to the Talmud, is "as if it were simply water." Embryos from which stem cells are derived develop only about three to five days in petri dishes. Therefore, conservative Jewish rabbi and philosopher Elliot Dorff authoritatively concludes that "frozen embryos originally created for purposes of overcoming infertility may be discarded (presumably after the couple has had as many children as they plan to produce or has given up in that effort), but they may also be used for good purposes. One such purpose is to produce stem cells for medical research. Indeed, couples should be encouraged to donate their extra embryos for such efforts." He further notes of embryos created outside the body that "[s]ince they cannot become a human being outside a woman's uterus, their status is even less than that of an embryo in the first forty days of gestation."[72]

According to University of Virginia professor of religious studies Abdulaziz Sachedina, Islamic tradition believes that a fetus gains full status as person at four months of gestation.[73] Regarding the moral status of early-stage embryos in laboratory petri dishes, Muzammil Siddiqi, former president of the Islamic Society of North America, says, "Our answer is that the embryo in this stage is not human. It is not in its natural environment, the womb. If it is not placed in the womb it will not survive and it will not become a human being. So there is nothing wrong in doing this [stem cell] research, especially if this research has a potential to cure diseases."[74]

Of course, this brief survey of the views of the Abrahamic religions does not exhaust the range of ethical thinking inspired by faith about the moral status of human embryos and stem cells, but it does make clear that sincere people of faith can reach very different moral conclusions about this research. If you are a member of a faith community in which God or his trusted representative tells you that an embryo is a full person deserving of all the protections we accord adult people, then when it comes to whether you're going to believe Ronald Bailey's views on stem cell research or what you think are God's, I know the answer. The problem is that if one set of believers accepts as true the proposition that dismantling an embryo for stem cells is the moral equivalent of yanking organs out of a baby, and others, equally faithful, believe that they are commanded to use stem cells from embryos to heal the sick, there is no obvious way to resolve the issue. It turns into a kind of recurring divine "He said, He said."

At the risk of being presumptuous, I still hope that the light science sheds on this issue will lead faith communities that oppose human embryonic stem cell research to come to a different understanding one day. In the meantime, I hope that other readers not so persuaded by their faith will find the following arguments enlightening and useful.

WHAT'S IN THE DISH?

Setting aside the conflicting religious views of the moral status of embryos, how might science and philosophy address the question posed by University of Pennsylvania bioethicists Arthur Caplan and Glenn McGee: "What's in the dish?"[75]

The embryos from which stem cells are derived have developed from three to seven days and are made up of around one hundred or so cells.

Embryos at this stage are barely visible to the naked eye. Peering through a microscope, one would see a ball of cells called a trophoblast surrounding the inner cell mass. The cells of the trophoblast would form the placenta and umbilical cord of a developing embryo, while the inner cell mass would produce all the cells of the body. Stem cells are derived from the undifferentiated cells found in the inner cell mass of embryos. Researchers remove the inner cell mass and grow them in culture in petri dishes, where they constantly renew themselves and remain undifferentiated. Researchers have now coaxed stem cells into becoming skin, liver, heart, muscle, nerve, and other kinds of cells.

The arguments of Robert George, a professor of jurisprudence at Princeton University and a member of the President's Council on Bioethics, and Patrick Lee, a professor of philosophy at the Franciscan University of Steubenville, are fairly representative of those made by other bioconservative opponents of stem cell research. They maintain: "The facts of science . . . are clear: Human embryos are not mere clumps of cells, but are living, distinct human organisms, the same as you and I were at earlier stages of our lives."[76] George and Lee argue that these facts mean that embryos are already human beings and therefore "[i]t is wrong to kill and dismember them—at any stage of their existence—in the hope of benefiting others."[77] They therefore accord full moral standing to cells that, unlike Judson Somerville, have no brains, no thoughts, no hopes, no feelings, and no expectations. But there are good reasons to doubt that their claims about the moral status of embryos are dispositive.

"The fact that every person began life as an embryo does not prove that embryos are persons," says Michael Sandel, a professor of government at Harvard University and also a member of the President's Council on Bioethics.

> Consider an analogy: Although every oak tree was once an acorn, it does not follow that acorns are oak trees, or that I should treat the loss of an acorn eaten by a squirrel in my front yard as the same kind of loss as the death of an oak tree felled by a storm. Despite their developmental continuity, acorns and oak trees are different kinds of things. So are human

> embryos and human beings. Sentient creatures make claims on us that nonsentient ones do not; beings capable of experience and consciousness make higher claims still. Human life develops by degrees.[78]

Let's take Sandel's point about the moral claims of sentient creatures a bit further. Insights about how we define the end of human life shed considerable light on what constitutes its beginning. We used to declare people dead when their hearts stopped beating, until new biomedical advances forced us to think more carefully about what we mean when we say someone is "dead." We now know that a body may still breathe, its heart beat, and its stomach digest, but that the person whose body it was is well and truly "brain dead." Thus, technological advance has focused our attention on the central fact that we exist only if our brains are still working. If our brain activity ceases—if our thoughts, memories, emotions, and intentions come to an end—we have ceased to be.

Therefore, it is widely agreed that it is permissible to dissect brain-dead bodies for their spare parts. In fact, donating one's organs via one's will is generally regarded as a moral act, worthy of great approbation. Like brain-dead bodies, embryos consisting of one hundred or so cells have no minds, memories, or intentions. Indeed, they have no brains, nor even any nerve cells. Consequently, since we do define "persons" as the sort of entities that do have brains capable of sustaining a mind, embryos clearly don't qualify. Indeed, there is another similarity: Both the brain-dead and embryos need the aid of people outside themselves in order to remain animate. One needs an intensive care unit and the other needs a willing womb.

Science shows us as well that embryos are not, contra George and Lee, "distinct" organisms. Take the easy case of identical twins. Since they develop from the same fertilized egg, their genes are identical. They clearly become "distinct" sometime after conception. What about the case of human chimeras? Human chimeras occur naturally when two eggs become fertilized but, instead of developing into nonidentical twins, they fuse in the womb, making a single individual with two distinct sets of genes. Or what about the increasingly common procedure of preimplantation genetic testing of embryos produced using in vitro fertilization in which one cell is taken from a two-cell embryo for genetic testing and the other, if implanted in the womb, can develop into a baby. Can it be said that preimplantation testing kills a twin?

What's interesting about both twins and chimeras is that they point clearly to the fact that who we regard as individuals does not depend on "coming to be at conception" as Lee and George maintain, since both twins and chimeras as individuals clearly "come to be" sometime after conception. What human twins and chimeras point to is the fact that individuals are bodies and brains, not cells that can mix and match and become more than one individual or fewer than two. Twins also point up the fact that having your own unique genome is not what makes you an individual; having your own brain and body does.

IS HEAVEN POPULATED CHIEFLY BY THE SOULS OF EMBRYOS?

And what are we to think about the fact that Nature (and, for believers, Nature's God) profligately creates and destroys human embryos? John Opitz, a professor of pediatrics, human genetics, and obstetrics and gynecology at the University of Utah, testified before the President's Council on Bioethics that between 60 to 80 percent of all naturally conceived embryos are simply flushed out in women's menstrual flows unnoticed. Keep in mind that this is not miscarriage we're talking about. The women and their partners never even know that conception has taken place; the embryos simply depart the womb in the menstrual flow. In fact, Opitz stated that embryologists estimate that *the rate of natural loss for embryos that have developed for seven days or more is 60 percent.* The total rate of natural loss of human embryos increases to at least 80 percent if one counts from the moment of conception. About half of the embryos lost are abnormal, but half are normal embryos that, had they implanted, might have developed into healthy babies.[79]

The plain fact is that millions of viable human embryos each year produced via normal conception fail to implant and are shed in women's normal menstrual flow. Does this mean the world is suffering a veritable holocaust of annihilated innocent human life?

But given Lee and George's insistence that every embryo is "already a human being," does that mean that if we could detect unimplanted embryos as they leave the womb, we would then have a duty to rescue them and try to implant them anyway? "If the embryo loss that accompanies natural procreation were the moral equivalent of infant death, then pregnancy would have to be regarded as a public health crisis of epidemic proportions: alleviating natural embryo loss would be a more urgent moral cause than abortion, in vitro fertilization, and stem-cell research combined," notes Sandel.[80]

As far as I know, bioconservatives such as George do not advocate the rescue of naturally conceived unimplanted embryos. But why not? In right-to-life terms, normal unimplanted embryos are the moral equivalents of a thirty-year-old mother of three children.

Of course, culturally we do not mourn the deaths of these millions of embryos as we would the death of a child—and reasonably so, because we do in fact know that these embryos *are not people*. Try this thought experiment: A fire breaks out in a fertility clinic and you can save either a three-year-old child or a petri dish containing ten seven-day-old embryos. Which do you choose to rescue?

Stepping onto dangerous theological ground, it seems that if human embryos consisting of one hundred cells or less are the moral equivalents of a normal adult, then religious believers must accept that such embryos share all of the attributes of a human being, including the possession of an immortal soul. So even if we generously exclude all the naturally conceived abnormal embryos—presuming, for the sake of theological argument, that imperfections in their gene expression have somehow blocked the installation of a soul—that would still mean that perhaps 40 percent of all the residents of heaven were never born; never developed brains; and never had thoughts, emotions, experiences, hopes, dreams, or desires.

Yet millions of intelligent people of goodwill maintain that seven-day-old embryos have the exact same moral standing as do readers of this book. Acting on this sincere belief, they are trying to block biomedical research on human embryonic stem cells that is desired by a majority of their fellow citizens.

Is there a way out of this politico-theological impasse? The President's Council on Bioethics held an extraordinarily interesting session on December 3, 2004, in which two different avenues for obtaining human embryonic stem cells were proposed in ways that might skirt right-to-life moral objections.

First, Howard Zucker and Donald Landry, two medical professors at Columbia University, proposed "a new definition of death for the human organism, an organism in development, and that is the irreversible arrest of cell division."[81] They pointed out that a good percentage of in vitro fertilized embryos consist of a mixture of cells, some containing the wrong number of chromosomes (aneuploidy), some with the normal number. Embryos with such cell mixtures often cease development by cell division and thus cannot develop into fetuses, much less babies. Zucker and Landry argue that such embryos can be considered dead, and the normal embryonic cells they contain can be harvested just as organs can be ethically harvested from brain-dead adults. (Animal experiments have already shown that normal cells harvested from defective embryos will produce normal tissues.) Thus, we get stem cells from an entity that could not, under any circumstances, have become a human being.

William Hurlbut, a consulting professor in the Program of Human Biology at Stanford University and another member of the President's Council on Bioethics, proposed alternative means of producing cloned human embryonic stem cells that right-to-lifers should not find morally objectionable. Hurlbut cited work by researcher Janet Rossant at Mount Sinai Hospital in Toronto in which she inactivated the cdx2 gene in mice. Once the cdx2 gene is inactivated, the mouse embryo cannot form a trophoblast—the tissues that grow into the placenta. However, embryonic stem cells do develop, although they cannot form an embryo. Hurlbut proposed an attempt to find similar genes that could be inactivated in the nuclei of adult human cells before they are installed in enucleated human eggs to produce cloned embryonic stem cells that are a genetic match for the person who donates the adult nucleus. (Transplanted cells and tissues produced by such therapeutic cloning would not be rejected by the donor's immune system.) Once the stem cells have been derived, the inactivated genes could be reactivated so that the stem cells could be used to produce normal transplantable cells and tissues.

"This process does not involve the creation of an embryo that is then altered to transform it into a non-embryonic entity," explains Hurlbut. "Rather the proposed genetic alteration is accomplished *ab initio*, the entity is brought into existence with a genetic structure insufficient to generate a human embryo."[82] Although the chairman of the President's Council on Bioethics apparently favors Hurlbut's proposal, I am not at all sure that it will satisfy the concerns of pro-life advocates. Some might object that Hurlbut's proposal doesn't differ all that much from the case in which crucial genes for the development of human brains were disabled at the single-cell stage to create anacephalic fetuses from which tissues and organs could be harvested. Of course, the main and decisive moral difference is that an anacephalic fetus would have to be carried for some weeks in a woman's womb while stem cells under Hurlbut's proposal would be derived from petri dishes.

Would using these techniques to derive stem cells reduce the number of embryonic souls populating heaven? Who knows? But these options may offer a possible way around the moral blockades that impede promising biomedical research on human embryonic stem cells. Should we halt current human embryonic stem cell research while these possible new avenues of research are being explored? Absolutely not. That would be surrendering to the moral bullying of a minority that wants to halt promising medical research that could cure millions on theological grounds that many of their fellow citizens do not share. As three researchers associated with the Harvard Stem Cell Institute responded to Hurlbut's proposal in the *New England Journal of Medicine*: "Hurlbut's proposal is a distraction from the central issue, which is whether it is morally justifiable to use preimplantation-stage human embryos in the search to understand and cure serious diseases. We believe it is justified, and the diversion resources to alternative approaches that offer no scientific benefit merely diminishes the likelihood of success."[83]

STEM CELLS, HOLD THE EGGS

The next step in stem cell research will occur when biotechnologists learn how to strip off the suppressing proteins from a mature cell's genes and transform it directly into a stem cell without having to use enucleated human eggs, taking eggs and all their accompanying politico-ethical difficulties out of the discussion. "It may eventually become possible to take a cell from any one of our organs and to expose it to the right set of environmental stimuli and to encourage that cell to return to a more primitive stage in the hierarchy of stem cells," explains former NIH director Varmus. "Under those conditions, one might in fact generate the cell with as great a potential as a pluripotent cell [capable of becoming many different, but not all, types of tissues] from a very mature cell. One might even in fact imagine generating a cell that is totipotent [able to develop into a complete organism under the right circumstances] in that manner."[84]

Stem cells produced this way would be identical to the human embryonic stem cells that currently must be harvested from embryos. A cell whose methylation and suppressor and promotor proteins have been stripped off could become a nerve stem cell, a liver stem cell, or a baby—depending on the intentions of the patients and doctors. As noted earlier, researchers are experimenting right now to see if new embryonic stem cells could be formed by introducing the nucleus of an adult cell into an already existing enucleated embryonic stem cell, with no eggs involved.

As we have seen, one day it may be possible to take any adult stem cell back to the embryonic, and hence protean, stage. But the research to figure out how to do that depends on work with embryonic cells, and the resulting cells, of course, would themselves be embryonic. So it turns out that people who oppose stem cell research on the ground that any cell that can *become* a human being already *is* a human being are essentially arguing that every cell in your body is another person.

"What happens when a skin cell turns into a totipotent stem cell [a cell capable of developing into a complete organism] is that a few of its genetic switches are turned on and others turned off," wrote Oxford Univeristy bioethicist Julian Savulescu in 1999. "To say it doesn't have the potential to be a human being until its nucleus is placed in the egg cytoplasm is like saying my car does not have the potential to get me from Melbourne to Sydney unless the key is turned in the ignition."[85] Since nearly every cell in the human body contains the complete genetic code of an individual, it is logically possible, using biotechnology, to turn every one of a person's cells into a complete new human being. If one doesn't turn on the ignition of a car (or doesn't strip the suppressor proteins from a nucleus and put the cell into a womb), then the car won't go (or the skin cell won't grow into a human being). In other words, simply starting a human egg on a particular path, either through fertilization or cloning, is a necessary condition for developing a human being, but it isn't sufficient. A range of other conditions must also be present.

"I cannot see any intrinsic morally significant difference between a mature skin cell, the totipotent stem cell derived from it, and a fertilised egg," said Savulescu. "They are all cells which could give rise to a person if certain conditions obtained." Those conditions include the availability of a suitable environment like a woman's womb. A petri dish is not enough. "If all our cells could be persons, then we cannot appeal to the fact that an embryo could be a person to justify the special treatment we give it," Savulescu concluded. "Cloning forces us to abandon the old arguments supporting special treatment for fertilised eggs."[86]

The DNA content of a skin cell, a stem cell, and a fertilized egg are exactly the same. The difference between what they are and what they could become is the environment in which their DNA is found. Thus, according to Savulescu, the mere existence of human DNA in a cell cannot be the source of a relevant moral difference. The differences among these cells are a result of how the genes in each are expressed, and that expression depends largely on which RNAs and proteins suppress or promote which genes. Does moral relevance really depend on the presence of the appropriate proteins in a cell? Trying to base moral distinctions on this level of biochemistry seems a bit quixotic.

In 1999 Edward Furton, who works at the National Catholic Bioethics Center, asked the National Bioethics Advisory Commission to remember that "[a]s a result of the tainted origin, many Americans who have deeply held moral objections to embryo destruction may choose not to receive any benefits from this new research."[87] No one is suggesting that people should be forced to use medicines that they find morally objectionable. Perhaps someday different treatment regimens will be available to accommodate the different values and beliefs held by patients. One can imagine one medicine for Christian Scientists (minimal recourse to antibiotics, etc.), another for Jehovah's Witnesses (no use of blood products or blood transfusions), yet another for Roman Catholics (no use of treatments derived from human embryonic stem cells), and one for those who wish to take the fullest advantage of all biomedical discoveries.

New medical technologies have often been opposed on allegedly moral and religious grounds. For centuries autopsies were prohibited as sinful. People rioted against smallpox vaccinations and opposed the pasteurization of milk. Others wanted to ban anesthesia for childbirth because the Bible declared that after the Fall, God told Eve, "In sorrow thou shalt bring forth children" (Gen. 3:16). Many objected to heart transplants as benefiting one person by murdering and stealing another person's organs. In the field of reproductive medicine, it was forbidden to ship condoms through the US Mail, and contraceptive pills were illegal in several states until the late 1960s. Many of the same people who oppose therapeutic cloning, such as the chairman of the President's Council on Bioethics, Leon Kass, also opposed in vitro fertilization, which, since the 1970s has enabled hundreds of thousands of infertile couples to have families and is now widely approved of by most Americans. Opposition to stem cell research is just the latest manifestation of this all-too-human fear of the unknown, and can doubtless be expected to fade, as has such earlier opposition to sensible medical research.

For the sake of Hwang Mi-soon, Judson Somerville, and millions (including about 250,000 Americans) who live with spinal cord injuries, let us hope the South Korean research into umbilical stem cells can soon be reliably replicated and improved upon. However, only more research will tell us whether the promise of adult umbilical cord and embryonic stem cells will be fulfilled. In late May 2005, Korean scientists published research in *Science* showing that they had succeeded in creating eleven cloned human embryonic stem cell lines that are genetic matches for eleven different patients. This breakthrough proves the concept of making cloned human embryonic stem cells that are perfect transplants for individual patients. Clearly, various lines of research should be pursued simultaneously to give us the best chance of discovering effective future treatments. It may well turn out that adult stem cells are good treatments for certain diseases, umbilical cord stem cells work best for others, and embryonic stem cells are better at curing still different maladies. Contrary to the claims of bioconservatives, it is not *either* adult and umbilical cord stem cells *or* embryonic ones; for the sake of millions of suffering patients, it's necessary to forge ahead on all three fronts.

CHAPTER 4

WHO'S AFRAID OF HUMAN CLONING?

The day after Christmas in 2002, Brigitte Boisselier, president of Clonaid, the reproductive cloning company founded by the Raelian UFO religious sect, shocked the world.

Standing before massed television cameras, Boisselier announced that the world's first human clone—code name: "Eve"—had been born.[1] The seven-pound baby was allegedly a clone of a thirty-one-year-old American woman whose partner was infertile. The Raelians claimed that four more baby clones would be born early in 2003.

The Raelian announcement of a daring and unprecedented advance in human power over nature naturally provoked immediate bioethical hand-wringing and a predictable call for a ban on the practice of reproductive cloning. "Today's announcement, if it is true, is alarming and disturbing," declared Leon Kass. "We offered reasons and arguments that go far beyond concerns about safety, and that add up to a deep and permanent objection to what the Raelians now claim to have done."[2] They didn't have to wait until someone claimed to have actually *done* it to get in a panic—the council had earlier issued a report concluding that cloning to produce babies would be "thoroughly unethical and should be outlawed."[3]

Of course, most knowledgeable people immediately suspected that the "birth" of Eve was a hoax. And so it turned out to be. Rael, the former French race car driver who founded the sect after an alleged meeting with space aliens, later boasted about the escapade: "Some media experts say we got between $600 million and $700 million worth of coverage and I did nothing."[4]

Claims to have produced the first human baby by cloning continue to surface from time to time. Maverick fertility specialist Panos Zavos of the Andrology Institute of America in Lexington, Kentucky, announced in London in January 2004 that he had impregnated an infertile thirty-five-year-old woman with a cloned embryo. The embryo was allegedly a clone of the woman's husband.[5] Tellingly, Zavos made no subsequent announcements about the *birth* of a clone nine months later.

Zavos created another stir in late August 2004 when he announced, again in London, that he had injected cow eggs with genetic material from three corpses to create hybrid cow-human embryos.[6] Using this technique, he claimed to have created embryonic clones of an eleven-year-old girl and a thirty-three-year-old man, both of whom died in car accidents, and an eighteenth-month-old boy who died during surgery. So far, the only things Zavos has actually been able to clone are sensational newspaper headlines.

This cloning craze all began with the birth of an unassuming little lamb named Dolly. In 1997 Scottish biotechnologist Ian Wilmut of the Roslin Institute near Edinburgh took a cell from the udder of a six-year-old sheep, added its genes to a hollowed-out egg from another sheep, and placed the egg in the womb of a third sheep. This process is called somatic cell nuclear transfer, (SCNT) because one is transferring the nucleus of somatic cells—any cell of a multicellular organism that does not participate in the production of reproductive or germ line cells—into an egg whose own nucleus has been removed.

This resulted in the birth of an identical twin sheep that was six years younger than its sister. This event was quickly followed up by the announcement that some Oregon scientists had cloned monkeys using embryonic cells.[7] Shortly thereafter came cloned mice, cows, goats, mules, and cats. Researchers quickly recognized that in principle it should be possible to clone humans. That prospect initially spooked a lot of people, and quite a few of them began calling for national and international regulators to ban human reproductive cloning.

In 1997, a week after the birth of Dolly was revealed, President Clinton rushed to ban federal funding of human cloning research and asked privately funded researchers to stop such research until the National Bioethics Advisory Commission could consider the matter. The National Bioethics Advisory Commission, composed of scientists, lawyers, and ethicists, had been created to advise the federal government on the ethical questions posed by biotechnology research and new medical therapies. Clinton asked the Commission to give him a report on the ethical implications of human cloning within in ninety days.

But Sen. Christopher Bond (R-MO) didn't wait around for the commission's recommendations. Bond introduced a bill right away to ban the federal funding of human cloning or human cloning research. "I want to send a clear signal," he said, "that this is something we cannot and should not tolerate. This type of research on humans is morally reprehensible."[8] Within ten days after Wilmut revealed Dolly's existence, three bills had been introduced in the US Congress to prohibit the cloning of humans and to outlaw federal funding for research in human cloning.[9]

Carl Feldbaum, president of the Biotechnology Industry Organization, hurriedly declared that human cloning should be immediately banned. Perennial bioluddite Jeremy Rifkin grandly pronounced that cloning "throws every convention, every historical tradition, up for grabs."[10] At the putative opposite end of the political spectrum, conservative columnist George Will chimed in: "What if the great given—a human being is a product of the union of a man and woman—is no longer a given?"[11]

In addition to these pundits and politicians, a whole raft of professional bioethicists leapt into the fray to declare that they, too, opposed human cloning. Daniel Callahan said, "The message must be simple and decisive: The human species doesn't need cloning."[12] George Annas agreed: "Most people who have thought about this believe it is not a reasonable use and should not be allowed."[13]

In early June 1997 the National Bioethics Advisory Commission recommended that human cloning be banned by federal legislation for up to five years, after which the ban would expire unless reenacted. Commission member Alexander Capron, a bioethicist at the University of Southern California, noted if such a ban were passed, it would be the first time a field of biomedical research had been outlawed.[14]

Since Dolly's arrival in 1997, the debate over the morality of reproductive cloning has never flagged. It flares up whenever a cloning hoaxer like Clonaid or Zavos makes a claim, or whenever progress toward creating cloned human embryos for biomedical research is reported. For example, when Advanced Cell Technology announced in November 2001 (prematurely, it turns out) that they had succeeded in getting a cloned human embryo to start development,[15] President Bush immediately, if incoherently, denounced it, saying, "The use of embryos to clone is wrong."[16] His Bioethics Council agreed: in 2002 it unanimously called for a ban on reproductive cloning covering "everyone, corporations as well as individuals, private as well as public institutions."[17] Opposition most recently erupted again when South Korean researchers reported in February 2004 that they had created thirty cloned human embryos.

Given the constant cries from so many corners of American officialdom, scientific and political, one would think it was crystal clear that cloning humans is unethical. But what exactly is wrong with it?

Which ethical principle does cloning violate? Stealing? Lying? Coveting? Murdering? What? Most of the arguments against cloning amount to little more than a reformulation of the familiar refrain of Luddites everywhere: "If God had meant for man to fly, he would have given us wings. And if God had meant for man to clone, he would have given us spores."

What would a clone be? He or she would simply be a complete human being who happens to share the same genes as another person. Today, we call such people identical twins. To my knowledge no one has argued that twins are immoral. Of course, cloned twins would not be the same age. But it is hard to see why this age difference might present an ethical problem, or give clones an inherently different moral status.

"You should treat all clones like you would treat all monozygous [identical] twins or triplets," concludes H. Tristram Engelhardt, a physician and philosopher at Rice University. "That's it."[18] It would be unethical to treat a human clone as anything other than a human being. If this principle is observed, he argues, all the other "ethical" problems for a secular society essentially disappear.

Let's take a sober look at the concerns human cloning opponents have raised. Even before the Raelians pulled off their "Eve" hoax, the *New York Times Magazine* published an article, "Lab of the Human Clones," about the group's purported efforts to produce a cloned baby. According to the *Times*, the Raelians were preying on a misconception that many people apparently hold: that cloning can somehow bring back or replace a dead child or other loved one. Of course, that's not possible. A clone would share the same genes as the dead loved one, but all the things that make us unique individuals are found in our brains; our loves, our memories, our characters, are irretrievably gone when we die. Genes code proteins, not memories or souls.

Others argue that clones would somehow undermine the uniqueness of each human being. "Can individuality, identity, and dignity be severed from genetic distinctiveness, and from belief in a person's open future?" asks Will.[19] The President's Council on Bioethics expresses the same concern in the overrefined diction characteristic of Kass's prose: "[O]ur genetic uniqueness is an important source of our sense of who we are and how we regard ourselves. It is an emblem of independence and individuality. It endows us with a sense of life as a never-before-enacted-possibility . . . we go forward as genetically unique individuals into relatively indeterminate futures."[20]

Are Will and the council suggesting that identical twins suffer from badly impaired senses of individuality and identity? They have apparently fallen under the sway of what former University of Virginia bioethicist John Fletcher called "genetic essentialism." Fletcher pointed out that polls indicate that some 30 to 40 percent of Americans are genetic essentialists who believe that genes almost completely determine who a person is.

But a person who is a clone would live in a very different world from that of his genetic predecessor. With greatly divergent experiences, their brains would be wired differently. After all, even identical twins who grow up together are separate people—distinct individuals with different personalities and certainly no lack of Will's "individuality, identity, and dignity." And because lives are lived by individuals, not genomes, all lives are fundamentally adventures in "never-before-enacted-possibility."

In addition, a clone that grew from one person's DNA inserted in another person's host egg would pick up "maternal factors" from the proteins and RNA in that egg, significantly altering his or her physical and mental development. Physiological differences between the womb of the original and host mothers could also affect the clone's development. In no sense would or could a clone be a carbon copy or a replica of his or her predecessor.

In any case, if ignorance of one's genetic heritage is crucial to our sense of identity and our belief in an open future, we'll soon all be in the same boat as future human clones. Why? Because of whole-genome genetic testing. Someday soon, University of Texas bioethicist John Robertson predicts, everyone will be able to go to his or her physician and obtain a complete readout of his or her genetic makeup for only one thousand dollars.[21] Such genetic profiles will give us all kinds of information about our disease risks and our psychological predilections. Yet having that kind of genetic information is far more likely to free us than it is to bind us. We would be able to take steps to avoid future heart disease or seek counseling to overcome innate shyness. Like all knowledge, genetic knowledge is power—the power to change and improve our lives, if we so choose.

All of which is not to say that some future cloned children's lives would not be negatively affected by the fact that they were born by means of cloning. Some parents of cloned children might treat them as means for their own self-aggrandizement. Consider the case of a rich jerk, so narcissistic that he wants to clone himself just so he can give all his wealth to . . . himself. First, he will fail. His clone is simply not the same person that he is. The clone may be a jerk, too, but he will be his own individual jerk. Nor is Jerk Sr.'s action unprecedented. Today rich people, and regular people, too, make an effort to pass along some wealth to their children when they die. People will their estates to their children not only because bonds of love connect them, but also because they have genetic ties. The principle is no different for clones.

The President's Council on Bioethics waxes grandiloquent about the mysteries of procreation: "Procreation is not making but the outgrowth of doing. A man and woman give themselves in love to each other, setting their projects aside in order to do just that. Yet a child results, arriving on its own, mysterious, independent, yet the fruit of the embrace."[22] The council does acknowledge in a sour footnote that "[w]e are, of course, well aware that many children are conceived in casual, loveless, or even brutal acts of sexual intercourse, including rape and incest."[23] Setting aside the flowery rhetoric, the fact is that about half of all births, for both married and unmarried women, are unplanned in developed countries today.[24] Children do result from a couple's embrace, and, if lucky, they are loved, planned or not. Given the considerable emotional and financial investment that the parents of cloned children will be making, I suspect that most future cloned children will be among the lucky ones, very much wanted and treasured by their families.

Bond and others worry about a gory scenario in which clones would be created to provide spare parts, such as organs that would not be rejected by the predecessor's immune system. "The creation of a human being should not be for spare parts or as a replacement," he declares.[25] If by "human being" Bond means an infant or adult clone, he is certainly correct. The simple response to this scenario is: People born as clones are people. You must treat them like people. We don't forcibly take organs from one twin and give them to the other. Why would we do that in the case of clones? On the other hand, if the senator is counting a one hundred-cell embryo as a human being, he is mistaken. It is not immoral to derive stem cells from one hundred-cell cloned embryos (see chapter 3).

The technology of cloning may well allow biotechnologists to develop animals that will grow human-compatible organs for transplant. Cloning is likely to be first used to create animals that produce valuable therapeutic hormones, enzymes, and proteins.

Another fear of the anticloners is a wave of obsessive cloning of exceptional human beings to get more of what they have. Will put it this way: "Suppose a cloned Michael Jordan, age 8, preferred violin to basketball? Is it imaginable? If so, would it be tolerable to the cloner?" Yes, it is imaginable, and the cloner would just have to put up with violin recitals. Kids are not commercial property—slavery was abolished some time ago. We all know about Little League fathers and stage mothers who push their kids, but given the stubborn nature of individuals, those parents rarely manage to make kids stick forever to something they hate. A ban on cloning won't abolish pushy parents.

One putatively scientific argument against cloning has been raised. As a National Public Radio commentator who opposes cloning once quipped, "Diversity isn't just politically correct, it's good science."[26] Sexual reproduction seems to have evolved for the purpose of staying ahead of ever-mutating pathogens in a continuing arms race. Novel combinations of genes created through sexual reproduction help immune systems devise defenses against rapidly evolving germs, viruses, and parasites. Thus, the worry goes, if enough human beings were cloned, pathogens would adapt and begin to get the upper hand, causing widespread disease. It will be, we are warned, like what happens when a lot of farmers all adopt the same corn hybrid. If the hybrid is highly susceptible to a particular bug, then the crop fails.

That warning may have some validity for cloned livestock, which may well have to live in environments protected from infectious disease. But it is unlikely that there will ever be millions of clones of any one person. Even when we master human cloning, genomic diversity will still be the rule for humanity. There might be more identical twins, triplets, and the like, but unless there are millions of clones of one person—and who would have the motive and the ability to get millions of women to bear them?—raging epidemics sweeping through hordes of human beings with identical genomes seem very unlikely.

But even if someday, by some weird possibility, millions of clones of one person did exist—who is to say that by then novel technologies wouldn't be able to control human pathogens? After all, it wasn't genetic diversity that caused typhoid, typhus, polio, or measles to all but disappear in the United States—it was modern sanitation and modern medicine.

So, am I saying we should just forge ahead with cloning human beings? No, not just yet. It's nowhere near safe enough. One of the chief problems in cloning mammals so far has been its inefficiency—it takes a lot of failed embryos to produce one healthy cloned animal. Right now, only a tiny percent of mammalian clones are long-term survivors, according to Massachusetts Institute of Technology biology professor Rudolf Jaenisch.[27]

Jaenisch also points out that many cloned mammal fetuses—for example, pigs, calves, and sheep—develop severe abnormalities. For example, during fetal development, some clones produce larger-than-normal placentas,[28] others are born twice as big as normal, and some are born with deadly anatomical flaws such as enlarged hearts or defective kidneys. What causes these problems?

Jaenisch's research indicates that the abnormalities in most cloned animals arise from a lack of correct parental genomic imprinting of their genomes. During the production of normal sperm and eggs, specific genes required for embryo development are shut down by means of chemical marks, or imprints. Some of these are in sperm, others in eggs. Only when sperm and eggs are combined do all of the crucial genes work together, allowing proper embryonic development. Genomic imprinting is seen only in placental mammals (like humans) and seems to be a result of the biological battle between the sexes. Males want their offspring to survive and so favor large progeny. Females who need to reserve nutritional resources for themselves tend to favor smaller progeny. Thus, imprinted paternal genes tend to promote fetal growth (such as fetal growth factor, Igf2) while maternal genes inhibit fetal growth (a gene known as H19). The balance of these factors determines the proper size of offspring.

To make a long story short, the problem with mammalian cloning is that in most cases the genes taken from the nuclei of mature somatic cells are not properly reprogrammed, as they are during the normal production of sperm and eggs. So when these mature somatic nuclei are inserted into enucleated eggs to produce embryos, their imprinting is most likely wrong. For example, gene expression analyses of tissues taken from cloned mice indicate that 4 to 5 percent of their overall genomes and 30 to 50 percent of imprinted genes are not correctly expressed.[29]

Since there's currently no way to restore proper imprinting, either paternal or maternal genes affecting fetal growth may end up being dominant, creating the developmental imbalances seen so far in cloned animals. "For cloning to be made safe, the parental genomes of a somatic donor cell would need to be physically separated and individually treated in 'oocyte-appropriate' and 'sperm-appropriate' ways," writes Jaenisch. "Such an approach is beyond our present abilities, implying that serious biologic barriers (rather than mere technical problems) hinder faithful reprogramming after nuclear transfer and thus preclude the use of nuclear cloning as a safe reproductive procedure."[30]

Perhaps Jaenisch is a bit too pessimistic about the prospects for safe human reproductive cloning. Tomohiro Kono and colleagues at the Tokyo University of Agriculture in Japan have overcome one of the technical problems posed by gamete imprinting. In April 2004 they announced the "virgin birth" of a mouse named Kaguya.[31] Kaguya has two mothers and no father. She was created by combining eggs from two different female mice—no males or sperm involved.

Normally, combining mammalian eggs this way won't work because of the imprinting problems described above. The Japanese researchers got around this problem by creating a female mouse genetically engineered so that her eggs behaved like sperm by turning on the paternally imprinted gene for promoting the production of Igf2 protein. The Japanese researchers then combined the genetically engineered mouse's eggs with the eggs from a normal mouse, installed the "reconstructed" eggs in surrogate mother mice, and Kaguya was born. The process is very inefficient, though. It took 457 reconstructed eggs to produce just two live baby mice. Nevertheless, this achievement shows that researchers are learning how to manipulate genomic imprinting, and that could lead to improved success in cloning mammals. The technical hurdles for successful cloning may be lower than many think. One day another approach might be possible in which researchers test a whole suite of pre-embryos and then implant only the ones with the proper imprinting.

For the time being research should focus chiefly on cloning other species until scientists understand why the abnormalities occur and can reliably correct them. For example, biologist Gerald Schatten at the University of Pittsburgh reported in October 2004 that he and his colleagues, using techniques devised by the Korean researchers who created human clones for stem cells in February 2004, were making progress on cloning monkeys.[32] So far, none of the pregnancies using cloned monkey embryos has lasted more than a month, but Schatten plans to continue his research. Just a year earlier Schatten speculated in *Science* that cloning primates might be impossible.[33]

A good benchmark for deciding to proceed with human reproductive cloning would be when researchers are reasonably sure that clones would suffer no more likelihood of birth defects (about 2 percent) than do children produced by sexual reproduction, either in vitro or by conventional means. It's too early now.

Proponents of fast-track efforts at human reproductive cloning might argue that such attempts are little different from the first efforts at in vitro fertilization, in which researchers implanted scores of embryos before one worked, bringing Louise Joy Brown, the first "test tube" baby, into the world. However, in the case of in vitro fertilization, animal research had identified no problems with increased birth defects resulting from the procedure. That's clearly not the case with mammalian cloning right now.

Should the federal government and other governments then outlaw attempts at human reproductive cloning? Attempting to clone a human being now would be incredibly negligent. It is appropriate to protect people from very negligent behaviors. However, bans have a way of becoming permanent. Nevertheless, if today's hasty cloning efforts were to produce a wave of defective babies, it would be one of the worst things that could happen to the hopes of infertile people who might look to reproductive cloning as way of having genetically related children. Public revulsion could then set back reproductive cloning for decades.

Still, in the long run there's no reason to think that a law against reproductive cloning would make much difference. “It’s such a simple technology, it won’t be bannable,” says Engelhardt. “That’s why God made offshore islands, so that anybody who wants to do it can have it done.”[34] As the cases of renegade would-be cloners Clonaid and Zavos already show, cloning would simply go underground and be practiced without legal oversight. Then people who turned to cloning would not have recourse to the law to enforce contracts, ensure proper standards, and hold practitioners liable for malpractice.

The chief problem with achieving and imposing a moral consensus on cloning—and many other biotech matters—is that Americans live in a large number of disparate moral communities. “If you call up the Pope in Rome, do you think he’ll hesitate?” asks Engelhardt. “He’ll say, ‘No, that’s not the way that Christians reproduce.’ And if you live Christianity of a Roman Catholic sort, that’ll be a good enough answer. And if you’re fully secular, it won’t be a relevant answer at all.”[35]

As we’ve seen, many prominent bioethicists claim that even safe reproductive cloning would be ethically wrong. As history shows, many bioethicists succumb to the thrill of exercising power by saying no. The bioethics field is littered with ill-advised bans, starting in the mid-1970s with the two-year international moratorium on recombining DNA, and also including existing laws against selling organs and blood and President Clinton’s prohibition on using human embryos in federally funded medical research.

Simply leaving people free to make their own mistakes will get a bioethicist no perks, no conferences, and no power. Ultimately, biotechnology is no different from any other technology—humans must be allowed to experiment with it in order to find its best uses and, yes, to make and learn from mistakes in using it. Trying to decide in advance how a technology should be used is futile. The smartest commission ever assembled simply doesn’t have the creativity of millions of human beings trying to live the best lives that they can by trying out and developing new technologies.

Once the public understands more completely the limitations of cloning—for example, that it really isn’t a way to bring back the dead—

human cloning will likely be used mostly by infertile couples who have no other choice for bearing biologically related children. For the rest of us, producing children the old-fashioned way will remain a lot cheaper—and a lot more fun.

CHAPTER 5

HOORAY FOR DESIGNER BABIES!

In a near-future world, parents may well be able to spare their children the agony of diseases such as cystic fibrosis and muscular dystrophy. Even more wonderfully, they may be able to better ensure that their children will be smart, skilled in certain respects, and less prone to depression. As paradisiacal as this world might sound, a variety of influential thinkers and bioethicists want to keep us from it—by any means necessary.

The beginnings of this world are already upon us. A mother was able to spare her infant daughter the horror of losing her mind by age forty by means of genetic testing.[1] Specifically, a married thirty-year-old genetics counselor who will almost certainly suffer from early-onset Alzheimer's disease by age forty chose to test her embryos in vitro for the gene that causes the ailment. She then implanted into her womb only embryos without that disease gene.

The result was the birth of a healthy baby girl—one who will not suffer Alzheimer's in her forties. The mother in this case certainly knows what would face any child of hers born with the disease gene. Her father, a sister, and a brother have all already succumbed to early Alzheimer's.

To achieve this miracle, the mother used the services of the Reproductive Genetics Institute (RGI) in Chicago, a private fertility clinic that's a pioneer of this type of testing, called preimplantation genetic diagnosis (PGD). PGD is being used by more and more parents who want to avoid passing on devastating genetic diseases to their progeny. Diseases tested for include cystic fibrosis, Tay-Sachs, various familial cancers, early-onset Alzheimer's, sickle-cell disease, hemophilia, neurofibromatosis, muscular dystrophy, and Fanconi anemia.

Currently, PGD is limited to detecting at most five different genetic diseases at a time, both because of the limits of current markers and because the amount of DNA that can be tested is very limited—it all comes from a single cell. However, researchers at Molecular Staging, Inc. in Connecticut now offer whole-genome amplification kits.[2] Using such kits, physicians can make as much DNA as they need from just one embryonic cell, allowing them to conduct hundreds of simultaneous tests for disease genes and other genetic traits.[3] British fertility specialist Alan Handyside, a professor of developmental biology at the University of Leeds and scientific director of the Bridge Fertility Center in London, plans to start offering whole genome embryonic screening in 2005.

PGD is still most commonly used by couples at high risk of producing embryos with an abnormal number of chromosomes. For example, Down syndrome results when an embryo has three copies of chromosome 21, instead of the normal two. The risk of Down syndrome increases with a woman's age, rising from 1 in 725 at age thirty-two to 1 in 109 at age forty. Fertility specialists can take a single cell from an eight-cell embryo and tag genes specific to chromosome 21 with a fluorescent dye. If they see three glowing dots, they know that the embryo likely bears three copies of chromosome 21, and would suffer from Down syndrome if implanted and brought to term. PGD has so far made possible the birth of nearly fifteen hundred healthy children worldwide.

So physicians can now test for disease genes in either human eggs or early-stage embryos. But these tests are not cheap. The PGD at the clinic in Chicago costs $2,500, on top of the $7,500 fee for the in vitro fertilization procedure. Still, the price is low when compared to the pain and suffering avoided. And as the tests become more common, their cost is likely to drop substantially.

PGD allows parents to avoid the heartbreaking decisions that come from using postconception means such as chorionic villus sampling or amniocentesis to detect genetic defects in fetuses. In chorionic villus testing, physicians insert a needle to sample cells from the developing placenta beginning at twelve weeks' gestation. Similarly, amniocentesis samples the amniotic fluid that surrounds a fetus in the womb looking for cells shed by the fetus. The fetal cells from the villi or the amnion are tested for chromosomal anomalies. If problems are found, parents can then make a more informed decision as to whether to bring the fetus to term.

The Chicago mother was able to bear a healthy baby because of the enormous medical progress that has been made over the past thirty years in assisted human reproduction. The first "test tube" baby, Louise Joy Brown, was born in 1978 thanks to the efforts of British fertility researchers Robert Edwards and Patrick Steptoe. Since that breakthrough, perhaps as many as two million babies worldwide have been born to hundreds of thousands of happy parents by means of in vitro fertilization technologies.[4]

Today, 2.5 percent of babies born in the UK are conceived via in vitro fertilization. In Finland, it is 3.2 percent. In the past quarter century, in vitro fertilization technologies have advanced, enabling patients and physicians to use donor eggs, donor sperm, and surrogate mothers to have children. Controversially, in vitro fertilization technologies have also been used to help gay couples, single people, postmenopausal women, and dead men (using either previously frozen sperm or sperm taken post mortem from their testes) have children.

Other reproductive technologies include intracytoplasmic sperm injection (ICSI)—the insertion of a sperm into an egg to cause fertilization. This technique is used when the father's sperm can't breach the egg's protective outer coatings without help. Tens of thousands of children have been born using this technique.

Another advance is oocyte cytoplasmic transfer. Fertility doctors use this when they determine that the cytoplasm—the material outside the cell's nucleus—in an older patient's eggs is somehow deficient. Doctors take cytoplasm from a young woman's egg and inject it into the patient's egg, rejuvenating it. In cytoplasmic transfer, mitochondria, the energy-producing organelles from the younger eggs, are transferred.

Since those mitochondria have their own small set of genes, children born with donated mitochondria have a changed germline. This means that female children born using this technique will pass those donated mitochondrial genes to their descendants through their own eggs. In a sense, a child born using this technique has three genetic parents. Some eighteen healthy children were born using this technique, although one was born with a form of autism. In 2002 the FDA asked practitioners to stop using the procedure, claiming that the agency has jurisdiction over the use of human cells that have received transferred genetic material by means other than the union of gametes. The agency ordered that any further ooplasm transfer protocols should be done under investigational new drug (IND) exemptions and that an IND submission to the agency would be required to treat additional patients.

At the Medical University in Taiwan, researcher Chii-Ruey Tzeng has devised a technique similar to oocyte cytoplasmic transfer. He extracts mitochondria from a woman's own granulosa cells and injects three thousand of them into her eggs. These women have been unable to get pregnant using other in vitro fertilization techniques, but this new procedure dramatically boosts pregnancy rates from 6 percent to 35 percent.[5] One concern is that, unlike the case of oocyte cytoplasmic transfer, the mitochondria derived from cells other than eggs might not behave properly in the long run. Researchers in Britain have now applied for permission to transfer egg nuclei into other eggs whose nuclei have been removed in order to prevent children from inheriting diseases caused by mitochondrial defects in their mothers.

Thus, preimplementation genetic diagnosis (PGD) is just the first step toward allowing parents to enhance their children genetically. The next steps in reproductive technologies will be toward repairing genetic diseases and genetically enhancing existing human capacities. Instead of selecting healthy embryos, researchers believe that it will be possible to correct disease genes such as those for sickle cell anemia, muscular dystrophy, and Tay-Sachs in embryos, so that the resulting children will be born healthy. The work of Richard J. Bartley, of the National Institute of Arthritis and Musculoskeletal and Skin Diseases, and his colleagues on repairing a single mutation that causes Duchenne muscular dystrophy hints at how this might be done.[6]

Children with Duchenne muscular dystrophy produce a defective version of the muscle protein dystrophin, which leads to progressively weakening muscles. Typically they die by age twelve of respiratory failure. To repair the defective gene, Bartlett attached RNA sequences that matched DNA sequences near the point of the mutation. In between the RNA sequences, Bartlett inserted a short DNA segment of the normal dystrophin gene. Bartlett injected his RNA-DNA sequences into a golden retriever that suffered from muscular dystrophy. Like guided missiles, the RNA sequences sought out and bound to the matching DNA surrounding the exact location of the mutation on the X chromosome. Once the RNA-DNA complex was bound to the chromosome, the cell's own DNA repair mechanisms detected the mismatch between the normal DNA segment and the mutated segment and spliced in the normal segment to replace the mutation. Once repaired, the cells produced normal dystrophin.

Other researchers have done comparable work in repairing sickle-cell anemia genes. Of course, this is preliminary work, and it was done in whole animals. But a similar technique could repair defective genes in embryos or even single fertilized eggs. In fact, that should be easier than performing genetic therapies aimed at repairing genes in infants or adults, since the targeted gene repair needs to be made in only one or a few cells.

These so-called germline interventions would be permanent genetic changes made in the embryos, and they would be passed on to future generations. These techniques not only cure one baby, but all of that baby's potential descendants as well.

Any technology that can safely correct disease genes can also be used to introduce "enhanced" genes into embryos. Someday parents using biotechnology will be able to ensure that their children have stronger immune systems, higher intelligence, and more athletic bodies than they otherwise would have.

For example, possibly just over the horizon are artificial chromosomes containing genes that protect against HIV, diabetes, prostate and breast cancer, and Parkinson's disease, all of which could be introduced into a developing human embryo. Already, Chromos Molecular Systems, based in Vancouver, British Columbia, makes a mammalian artificial chromosome that allows biotechnologists to plug in new genes just as new computer chips can be plugged into a motherboard. These artificial chromosomes, which have been developed for both mice and humans, offer exquisite control over what genes will be introduced into an organism and how they will operate.[7] Artificial chromosomes carrying genes selected by parents could be inserted into an embryo at the one-cell stage. Once the artificial chromosomes were incorporated into the embryo's genome, the selected genes would spread normally. They would be in every cell of the enhanced child's body when he or she was born.

When born, a child carrying such an artificial chromosome could have, for example, a souped-up immune system. Even more remarkably, artificial chromosomes could be designed with "hooks" or "docking stations," so that new genetic upgrades could later be slotted into the chromosomes and expressed when the child becomes an adult. Artificial chromosomes could also be arranged to replicate only in somatic cells, which form only the tissues that make up the body, and not in the germ cells involved in reproduction. As a result, genetically enhanced parents would not pass those enhancements on to their children; they could choose new or different enhancements for their children, or have them born without the benefit of any new genetic technologies at all.

But many respected scientists believe that safe genetic engineering of people may never be practical. Most human traits—both desirable and deleterious—are the result of the actions of multiple genes. Consequently, they argue that efforts at genetic enhancement may fail due to the so-called combinatorial explosion.

A combinatorial explosion occurs when a huge number of possible permutations are created by increasing the number of entities that can be combined. For example, there are 86,493,225 ways to pull twelve rabbits out of a hat containing thirty rabbits, and more than 635 billion thirteen-card bridge hands that can be dealt from a fifty-two-card deck. The combinatorial problem grows mind-bogglingly huge when one considers the various ways just twenty to twenty-five thousand human genes and the one hundred thousand or so proteins they produce can be combined in human cells and tissues. So skeptics of genetic engineering argue that such a vast array of complicated interactions of genes, environmental stimuli, and cellular processes may well preclude safe and effective engineering in human beings for such multifactorial genetic illnesses as heart disease and cancer, and for beneficial traits such as high intelligence. But perhaps this is too dismal a long-term view.

Molecular biologists are in the very early stages of attacking this problem. Now that the human genome—the complete genetic recipe for making a human being—is mapped, scientists are beginning to probe the complex ecology and interactions of genes to see how the recipe actually works to create a human.

One promising approach is to use biochips for testing tissues to see which genes are turned on or turned off during disease states. Biochips look a lot like silicon electronic microchips. But instead of circuitry, the surface is lined with thousands of bits of DNA, which attach to complementary strands of DNA from samples when they match. The DNA bits on the surface of biochips are like hooks, each catching only one very specific DNA fish as it swims past. Biochips can also measure messenger RNA, the molecule used by genes to tell cells what proteins and other substances to make. By detecting which messenger RNAs are present, researchers are able to tell which genes are active in a cell or tissue.

This kind of testing provides preliminary information about how genes interact to produce health or disease in various tissues. It also shows how old cells differ from young cells. By comparing diseased and healthy tissues on the gene level, researchers hope to identify gene and protein targets for pharmaceuticals that would restore diseased tissues to health.

Multiple gene testing also can identify the constellations of genes that make their bearers more susceptible to various diseases, such as heart disease or Alzheimer's. For example, a recent study tested for three genes that, when combined, improved by eightfold the ability to predict patients who are prone to dangerous blood clots, a condition known as venous thrombosis.[8] In the future, one can imagine confronting this deleterious genetic condition through both pharmaceutical intervention and engineering embryos with a more beneficial set of genes.

Such genetic engineering may not be as complicated as it sounds, since constellations of genes sort themselves into specific sets called haplotypes. Haplotypes are blocks of gene variants that travel down generations together. In a sense, evolution, by testing various combinations of genes and devising haplotypes, has already cut through the combinatorial explosion for us. If individual genes are like the separate ingredients for a cake—flour, baking soda, sugar, salt, yeast, and so forth—haplotypes are more like cake mixes in which all the ingredients are premixed in definite and predictable ways. So by using known haplotypes, future genetic engineers will not be reinventing the wheel. They will simply use already existing gene combinations, tested by evolution and time to enhance a future child's health or mental agility.

A first step in getting into the most intimate mechanics of the gene may come not from actually getting down and dirty in our smallest parts, but by designing exact electronic simulations. The "virtual cell" is another project whose results will be invaluable to future genetic engineers. Researchers such as Stanford's Harley McAdams and Lucy Shapiro are trying to build computer models of human cells, tissues, and organs to simulate complete genetic regulatory and metabolic pathways. Such models would be useful aids for testing pharmaceuticals "in silico." Virtual cell simulations could also help future genetic engineers predict how prospective genetic changes would cascade through cells and tissues, thus enabling them to avoid bad unintended consequences.

Safe human genetic engineering is still a ways off, but as the nineteenth-century British chemist Michael Faraday once declared, "Nothing is too wonderful to be true, if it be consistent with the laws of nature."[9]

SEX SELECTION

Parents are already able to use PGD to select some of the traits they want in their children. A political debate now facing us is over what limits, if any, should be placed on the traits parents are allowed to select. Sex is the easiest trait to select using PGD. Such attempts have a long history, from the herbal nostrums recommended by traditional healers to more recent therapists' advice about which forms of intercourse are allegedly likely to produce girls or boys. Sex selection is becoming a national crisis in India and China, where cheap mobile ultrasound clinics travel the countryside testing pregnant women. Women in these countries who discover their fetus is female often opt for legal abortions. The natural sex ratio is about 105 boys per 100 girls, but in India it is now 113 boys per 100 girls and as high as 156 boys per 100 girls in some regions. In China the sex ratio in recent years has been just shy of 120 boys per 100 girls. Such results have led both China and India to ban ultrasound testing for sex selection purposes.

In the United States, abortion is very rarely used for sex selection. A variety of new techniques do exist, though, that can help parents select their next child's sex. Consider flow cytometry, a method for sex selection in farm animals that fertility specialists have now adapted for human beings. This technology tags sperm bearing X chromosomes (those that determine females) and sperm bearing Y chromosomes (those that determine males) with a fluorescent dye so they can be segregated into different batches. The dye harmlessly attaches to the DNA molecules that make up genes. Female-determining X chromosomes are much bigger than male-determining Y chromosomes, which means that human sperm carrying X chromosomes have 2.8 percent more DNA than do sperm with Y chromosomes. Thus, sperm with X chromosomes soak up more of the fluorescent dye and glow more brightly. This difference in brightness allows flow cytometry machines to separate the X-bearing sperm from the Y-bearing sperm.

Once the sperm have been segregated, they may be used in either artificial insemination or in vitro fertilization to produce a child of the desired sex. Using sex-segregated sperm in artificial insemination also sidesteps the contentious debate over the moral status of embryos, since fertilization takes place directly in the would-be mothers' wombs.

This sperm segregation service has been developed by one of the world's leading centers offering artificial reproductive services, the Genetics and IVF Institute in Fairfax, Virginia. The institute's MicroSort service was first made available to patients in 1998. Ten of the first eleven babies born using it were girls. Fertility specialists refer to MicroSort as preconception sex selection (PSS). So far, more than five hundred babies have been born using MicroSort's methods for preconception sex selection.

Sperm segregation also helps prospective parents avoid giving birth to a male child who could suffer from one of the five hundred or so X-linked genetic diseases, such as hemophilia. Boys are particularly subject to these diseases because they inherit only a single X chromosome from their mothers, whereas girls inherit two, one from each parent. If there is a defective gene on one X chromosome, the undamaged one on the other X chromosome shields girls from its deleterious effects. But boys, who have only one X chromosome paired with a much smaller Y chromosome, will suffer from the disease. In a 2001 ethical statement, the American Society for Reproductive Medicine concluded that if PSS is found to be safe and effective, "physicians should be free to offer pre-conception gender selection in clinical settings to couples who are seeking gender variety [a child of the opposite sex to their first child] in their offspring."[10] The committee added some important caveats, including requirements that patients be fully informed of the risks of failure and that they give assurance that they will fully accept children of the opposite sex should the PSS fail.

And it might fail. The sperm-separating technique is not perfect: Batches of sperm intended to produce males typically contain 75 percent Y chromosome sperm, while the female batches contain 91 percent X chromosome sperm.[11] So the Genetics and IVF Institute and other fertility clinics now offer their patients a technique with a nearly 100 percent chance of producing a child of the desired sex: preimplantation embryo sex selection.

In preimplantation sex selection, prospective parents' eggs and sperm are combined in a petri dish, producing several embryos. Doctors then take a single cell from each embryo and test it to see which sex chromosome it bears. Only embryos of the desired sex are implanted in a woman's womb. The chances that this technique will result in a successful pregnancy that comes to term are similar to those of in vitro fertilization in general: about 20 to 30 percent.

Abortion opponents, who believe that embryos are people, naturally oppose this technique, because the embryos of the undesired sex will be destroyed. (Of course, fertility clinics regularly produce embryos that are never implanted and are often destroyed—or, now, could be used in embryonic stem cell research.) Setting aside those arguments, some bioethicists have other concerns about sex selection, either preconception or preimplantation.

The biggest such concern is a vision of a lopsidedly male planet, resulting in unhappiness and social unrest. However, unlike in China and India, polls show that Americans generally do not express a strong preference for children of either gender. In addition, the Genetics and IVF Institute's clinical data should soothe those worried about whether such sex selection technology will radically skew US sex ratios. Of the more than three thousand sperm-sorting cycles requested by patients, 77 percent have been seeking to produce girls. Thus, most ethicists agree that that is not a major issue here.[12]

The British activist group Human Genetics Alert argues against permitting parents to use sex selection technologies claiming that "the very act of choice is likely to harm the child."[13] How does sex selection allegedly harm a child? Human Genetics Alert asserts that parents' choices will be based on harmful sexual stereotypes and that children will become "just another human-designed consumer commodity, or object. The relationship becomes one between designer and object, where the latter is inevitably in a subordinate position."[14] But selecting an embryo because it is a girl or a boy doesn't obviously disadvantage the eventual child that develops. Surely Human Genetics Alert activists wouldn't deny that the experiences of raising boys and girls do differ and that most parents find those differing experiences fulfilling in different and equally valuable ways? Far from treating their sex selected offspring as mere commodities, it seems far more likely that parents who have gone to the trouble to take advantage of sex selection techniques will highly value the resulting child.

The worry that sex selection essentially regards one sex as inferior to another seems misplaced. Perhaps some parents in the United States and Europe harbor such bigoted notions, but this concern is somewhat allayed by the evidence that 90 percent of couples choosing sex selection are doing it because they wish to have children of both sexes in their families. "Sexism will no more be reinforced by PGS [pre-conception gender selection] or human cloning than it is now by pre- and postnatal gender stereotyping," insists Judith Daar, a bioethicist at Whittier Law School.[15] In other words, if sexism is a problem, we should address it in society at large, not parents' private reproductive choices.

Sex selection is the first example of genetic selection for a nondisease trait. Being a boy or girl is not a disease. Few aspects of human development are more significant than one's sex; it's a central fact of one's identity. If it is ethically permissible for parents to make that choice, the case for letting them make less significant genetic choices, including ones aimed at genetic enhancement, is already made.

THE ENEMIES OF GENETIC ENHANCEMENT

One might be tempted to dismiss concerns about genetic enhancement as premature. However, even bioconservatives like Francis Fukuyama don't think so. "As we discover not just correlations but actual molecular pathways between genes and traits like intelligence, aggression, sexual identity, criminality, alcoholism, and the like, it will inevitably occur to people that they can make use of this knowledge for particular social ends," worries Fukuyama in his book *Our Posthuman Future*. "This will play itself out as a series of ethical questions facing individual parents, and also as a political issue that may someday come to dominate politics."[16]

The prospect of a safe biotechnology that enables parents to enhance their children clearly frightens many prominent policy intellectuals on both the political Left and the Right.

The godfather of the antibiotech left is activist Jeremy Rifkin and his Foundation on Economic Trends. He is now ably assisted by environmentalist Bill McKibben. Also arguing against biotech from the left is George Annas, who proposes a global ban on reproductive cloning and all interventions in the human germline, including those aimed at curing genetic diseases.

But objections to biotech cross traditional political boundaries. As we've seen fighting against bioprogress from the right we have, most prominently, Leon Kass, head of the President's Council on Bioethics. He is joined by such leading conservative intellectuals as *Weekly Standard* editor William Kristol and Francis Fukuyama.

When grappling with the fears and anxieties of biotechnophobes, we should remember that we are still in many ways arguing about potentialities, not current realities. Enabling parents to genetically enhance their children is not going to be as easy as some of us might hope, nor will it happen as soon as we might wish. Right now nascent genetic enhancement technologies are simply not safe enough for use. Any future enhancement technologies will have to be thoroughly tested in animals before they can be used to help people.

Fortunately, our quickly advancing understanding of the complex web of interactions between genes and other cellular activities is likely to dramatically reduce the risks that might accompany inserting beneficial genes. A good general benchmark is that attempts at genetic enhancement should be delayed until solid research indicates that the risk of birth defects using such technologies is at least no greater than the risks of birth defects in children produced conventionally.

However, even those who fear their arrival agree that safe genetic enhancements are just over the horizon. Bioconservatives on both the Left and the Right fear that future biotechnological progress will transform humanity so much that our descendants will become, as Fukuyama says, "posthuman." They regularly invoke the dystopian visions of Mary Shelley's *Frankenstein*, Aldous Huxley's *Brave New World*, and C. S.

Lewis's *The Abolition of Man* as warnings of what a future of unleashed human biotechnology might hold. Fortunately, they are wrong.

Many opponents of human genetic engineering are either conscious or unconscious genetic determinists. They fear that the advance of biotechnological knowledge and practice will somehow undermine human freedom. In a sense, these genetic determinists believe that human freedom resides in the gaps of our knowledge of our genetic makeup. Typical in this regard is McKibben. Like other bioconservatives, McKibben accepts that the fondest dreams of the proponents of human genetic engineering eventually could come to pass. He even admits that advanced biomedical science could someday spare children from congenital diseases, cure cancers, correct disabilities, and lengthen the human life span.

Most people would embrace these possibilities with optimism, even joy. But for McKibben, these seeming advantages spell a dismal future for mankind. Parents who choose to use genetic engineering will end up turning their children into "robots" and "automatons." "Down that path," he declared in a recent debate, "lies the death of what we call human meaning, the idea that people are in some way their own human beings and are not pre-programmed semi-robots."[17]

Human freedom, McKibben evidently believes, depends in some profound sense on the random inheritance of the genes that are the recipes for our bodies and brains. As a result of this random genetic inheritance, he suggests, we have greater scope for freedom than if our genes had been chosen for us. McKibben is indulging in genetic essentialism: the unwarranted idea that we are just meat puppets dangling from our strands of DNA.

McKibben is obviously right when he declares, "[*G*]*enes do matter*."[18] But they don't matter as much as McKibben thinks they do. Take the case of monozygotic twins who share exactly the same genes and were formed in the same womb at the same time. They are certainly not identical people. In fact, variance between traits such as intelligence, personality, and even weight correlate only 60 to 70 percent between identical twins. That's much closer than with nonidentical siblings, but it's still a wide variance.

The case of identical twins proves the point that it is our brains, and not our genes, that make us individual human beings. That's why, in recent years, our society has legally defined death as brain death. Once our brains are gone, we are gone, even though our bodies—with all their genes—may live on. The fact is that we respect people, not their genomes. In a very real sense, we are no longer at the mercy of genes. Our genes are now at the mercy of our brains.

Biology also increasingly reveals that human individuality doesn't depend just on having different genes; it is the result of the interplay between genes and environment. A gene that enhances one's capacity for music doesn't mean that its possessor must become another Wolfgang Amadeus Mozart or Mick Jagger. Genes simply don't work that way. All of us have many capacities stemming from his or her specific genetic endowment. Perhaps I could have become a professional basketball player or a computer engineer, but I chose not to develop those particular abilities, despite the fact that my specific complement of genes might have allowed me to do so.

Genes order the production of different proteins in response to environmental influences such as schooling, physical training, infections, and nutrition. For example, as we learn, the genes in brain cells produce new proteins to fix memories and strengthen connections between nerves. Human genes are the necessary recipes for making human brains and bodies, but brains and bodies are manifestly shaped by their experiences. It might be possible someday, using genetic engineering, to give a child a brain smart enough to understand mathematician Andrew Wiles's proof of Fermat's Last Theorem. But she will have to undergo the experience of learning mathematics first. There are no genes for specific math problems.

Human freedom cannot and does not rely on ignorance and randomness. Human freedom—the capacity to make choices based on reason—expands with knowledge. If you don't believe it, think about how humanity's greater knowledge of such things as the germ theory of disease and the atomic theory of matter have radically increased humankind's choices and freedom during the last two centuries. Most of us would agree that there has certainly been an improvement over our ancestors' world—a world filled with friendly and hostile animistic spirits, one in which half of all children died before their first birthday.

Similarly, knowledge about how our genes affect our behavior and how our brains are wired increases rather than limits our freedom. Prozac, for example, does not limit our choices; it gives depressed people the freedom to adjust their emotional state. Ignorance is not freedom. Knowledge is freedom; ignorance is slavery.

In any case, if McKibben really believes that human freedom depends on inheriting a random selection of genes, his cause is already lost. Why? Genetic testing. Even McKibben recognizes that such testing will soon be here. "The biotech pioneer Craig Venter said in 2002 that within five years a personalized printout of an individual's genetic code would be cheap enough for anyone to buy, so you'll probably be able to afford it late next week or so," he writes.[19] In fact, in 2002 Venter offered to scan an individual's entire genome in a week for only $712,000.[20] Genetic testing will enable every one of us to know precisely our entire complement of randomly acquired genes. The good news is that we will then know our predispositions to various diseases, enabling us to take steps to delay their onset or even prevent them altogether.

To McKibben, such knowledge is a blow to our freedom, because we will also know a lot more about how our particular sets of genes influence our temperaments, our intelligence, our memories, and our physical capacities. Of course, that knowledge may well expand our freedom and our choices by making it possible for us to intervene by means of pharmaceuticals and optimized training to change our temperaments, improve our memories, or strengthen our bodies. Human freedom will then properly be seen as residing, at least partially, in our ability to overcome these predispositions, much as a former alcoholic can overcome his thirst for booze or an overweight person can lose weight by dieting.

McKibben's fears that genetic engineering will reduce human freedom are misplaced. To the extent that genes "program" us, we are already "preprogrammed" by our randomly conferred genes; we are just ignorant about which ones are doing what programming. But that won't be the case in the near future.

Providing children with such enhanced capacities as good health, stronger bodies, and cleverer brains, far from turning them into robots, will give them greater freedom and more choices. Almost everyone would want to have these beneficial traits. Those of us who regard a poor immune system, a weaker body, or an IQ of 80 as privations will welcome the opportunity to help our children avoid such conditions, even as we try now to keep our children safe and healthy, and to inspire and educate them.

McKibben objects that future gene-enhanced children will not have consented to receiving the genes selected by their parents. "The person left without any choice *at all* is the one you've engineered," he asserts. "You've decided, for once and for all, certain things about him: he'll have genes expressing proteins that send extra dopamine to alter his mood; he'll have genes expressing proteins to boost his memory; to shape his stature."[21]

To the extent that this is true, it is true for unengineered kids now. It's just that parents don't know which genes they've conferred on their children. Of course, they hope for the best—that their kids got the genes for good health, strong bodies, and sound brains. But there's always a chance they ended up with Grandma's genes for early heart disease or those that led to Uncle Jim's schizophrenia. Genetic engineering will help parents in the future avoid some of those harmful outcomes.

McKibben is correct that a gene-engineered child would have no choice about whether to express the proteins that lead to early onset Alzheimer's disease. But it's a pretty good bet that kids won't regret their parents' decision to eliminate those deleterious genes. But before we unthinkingly accept McKibben's misleading concerns about a child's informed consent, we should keep firmly in mind that *not one of us now living was asked our consent to be born, much less to be born with the complement of randomly conferred genes that we carry.*

Let's say a parent could choose genes that would guarantee her child a 20-point IQ boost. It is reasonable to presume that the child would be happy to consent to this enhancement of her capacities. How about plugging in genes that would boost her immune system and guarantee that she would never get colon cancer, Alzheimer's, AIDS, or the common cold? Again, it seems reasonable to assume consent. These enhancements are general capacities that any human being would reasonably want to have. In fact, lots of children already have these capacities naturally, so it's hard to see any moral justification for outlawing access to them for others.

Instead of submitting to the tyranny of nature's lottery, which cruelly deals out futures blighted with ill health, stunted mental abilities, and early death, parents would be able to open more possibilities for their children to lead fulfilling lives. Genetic enhancements to prevent these ills would not violate a child's liberty or autonomy, and certainly do not constitute a form of genetic slavery, as some opponents claim. Giving children such enhanced capacities as good health, stronger bodies, and cleverer brains, far from constraining them, would in fact give them greater freedom and more choices.

It's surely a strange kind of despotism that enlarges a person's abilities and options in life. In fact, through the gift of technology conjured from human intelligence, parents can increasingly bestow not only the gift of life, but also the gift of good health on their children. The good news is that any would-be tyrannical parents who buy into the bioconservatives' erroneous notions of hard genetic determinism will be disappointed. Their children will have minds and inclinations distinctly their own, albeit genetically enhanced.

In general, then, applying a reasonable-person standard, we can presume consent for general capacities that anyone would want, for instance, genes that tend to increase intelligence, strengthen immune systems, and lengthen lives.

But opponents of genetic enhancement try to frighten the public and policymakers by trivializing the choices that parents might make, implying a wave of genetic manipulations going far beyond the general beneficial enhancements we've been discussing so far. They suggest that some parents will want to genetically engineer piano prodigies or professional basketball players. Others raise the specter of Nazi eugenics by hinting that some parents will choose to endow their children with blond hair and blue eyes. Some have even suggested that black parents might choose to genetically engineer their children for white skin, in order for them to avoid the pain of racism.

It is doubtful that many parents will make such trifling choices if implementing them would pose even the slightest risk to their children's health. Engineering genes that improve a child's health may be worth some small risk, but trying to insert genes for red hair would likely not be. However, in order to avoid rare cases of parental misconduct, it may be necessary to limit by law genetic interventions to the types of general capacities described above.

Bioskeptics offer even more outlandishly science fiction nightmares to scare the public away from genetic engineering. Might not, they suggest darkly, parents choose to give their children gills or wings or other freakish genetic combinations? Even if such scenarios would be possible, they are unlikely. After all, human beings didn't sprout wings in order to fly, nor grow gills in order to swim underwater. Instead of modifying our genes, humanity will develop new technologies that will enable us to go where we want to go and do what we want to do.

Ultimately, applying a reasonable-person standard to genetic enhancements should allay the more lurid fears of biotech opponents and help citizens and policymakers as they craft whatever institutions are really needed to guide the safe development of human genetic engineering. But in doing so, they should not be guided by bizarre fears intended to keep humanity from empowering itself through better biotechnologies.

But McKibben, for one, thinks we can and should consciously refuse to so empower ourselves. Given all his worries, what does McKibben want us to do? He wants us to say "Enough" along with him and reject the Promethean prospects before us. Humanity should decide collectively to limit its technological questing once and for all. This is not an impossible dream, he thinks, because some societies have, at times, chosen to relinquish some technologies. The examples he wants us to follow, however, involve a pair of backward autocracies—fifteenth-century Ming China and seventeenth-century Tokugawa Japan—and the contemporary Amish, an example that actually undermines his argument.

Between 1405 and 1430 the Chinese admiral Zheng He made at least seven major voyages with the largest fleet the world had ever seen. These "treasure fleets" consisted of three hundred huge ships holding a troop of nearly thirty thousand people. Zheng He's fleets visited Java, Sumatra, Vietnam, Siam, Cambodia, the Philippines, Ceylon, Bangladesh, India, Yemen, Arabia, and Somalia. McKibben actually praises the Chinese emperors who chose to burn the great treasure fleet and destroy all records of the voyages. To prevent further adventuring, these emperors made it a capital offense to build a boat with more than two masts. Thus, declares an approving McKibben, "a great people turned its back on a promising technology." He adds, "The Chinese chose their definition of meaning—progress within tradition—over the pell-mell dynamism of the West."[22] But did a "great people" really choose to forgo the blessings of technology and trade? Isn't a far more reasonable interpretation that the rulers of China, who wanted nothing to disrupt their iron hold over the lives of their subjects, made that decision for the people?

Next, McKibben, without apparent irony, describes Tokugawa Japan as "a highly advanced feudal society." He praises it for outlawing firearms for two centuries. Why? Because "the samurai simply felt that guns were crude, that any peasant could use one," explains McKibben.[23] Which is precisely the point—naturally the beneficiaries of a feudal warrior society would want to make sure that the peasants didn't get hold of such equalizers. The peasants didn't relinquish firearms; their masters did it for them. But why worry about the meanings of the lives of Japanese peasants who were so downtrodden that they were forbidden the dignity of legal family names until after 1867? McKibben shows some disturbing colors by his enthusiastic approval of two societies in which technological progress was stifled for the benefit of absolute rulers.

The third case cited by McKibben, the Amish, is different and proves the opposite of what he thinks it does. The Amish live in an open society—ours—and can opt out of our society or theirs whenever they want. They have a system for voluntarily deciding among themselves what new technologies they will embrace. But the fact that they live as they wish and select only the technologies they want dramatically undercuts McKibben's point. The Amish case shows that technological choices don't have to involve everyone in a given society.

Like the Amish, bioconservative technophobes such as McKibben are free to say no to whatever technologies they wish to reject, for whatever peculiar reason. They do not have to genetically engineer their children or choose to live longer lives. But McKibben should be content to allow the rest of us to use those technologies we believe will enhance and improve our lives and the lives of our children. McKibben's mantra is always "More is not better." That's true, but it's completely beside the point: Better is better. And better, like beauty, is in the eye of the beholder.

BIOTECH AND ENLIGHTENMENT

McKibben has more specifically political concerns about bioengineering. He fears it will exacerbate inequality, even as he worries about homogenization. In his first scenario, the rich get access to safe genetic enhancements first, dramatically widening the gap between the rich and the poor. "The political equality enshrined in the Declaration of Independence can't withstand the destruction of the idea that humans are in fact equal," writes McKibben.[24] He is citing a similar worry expressed by Fukuyama, who declares, "The political equality enshrined in the Declaration of Independence rests on the empirical fact of natural human equality."[25] As noted earlier, Marcy Darnovsky, the associate executive director of the left-wing Center for Genetics and Society in Oakland, California, argues that supporters of biotechnological progress will violate the most sacred tenets of American democracy by "inscrib[ing] inequality onto the human genome."[26]

But are people "in fact equal"? There is nothing at all self-evident about physical human equality or equality of status. Some people are short, some tall; some fat, others thin; some strong, others weak; some poor, others rich; some brilliant, others dim. In other words, what we see is not self-evident equality, but human diversity and human individuality.

The ideal of political equality arose from the Enlightenment insistence that since no one has access to absolute truth, no one has a moral right to impose his or her values and beliefs on others. Political equality has never rested on claims about human biology. We all had the same human biology during the long millennia in which slavery, purdah, patriarchy, and aristocratic rule were social norms. Over the last two hundred years, human biology didn't change—our politics did.

The modern ideals of democracy and political equality are sustained chiefly by the insight, developed by Enlightenment thinkers, that people are responsible moral agents who can distinguish right from wrong, and therefore deserve equal consideration before the law and a respected place in our political community. The broad ability to distinguish right from wrong does not depend on the genetics of IQ, skin color, or gender. With respect to political equality, genetic differences are already differences that make no difference. Having some citizens who take advantage of genetic technologies and others who do not won't alter that principle.

Bioconservatives on both the Left and the Right also worry that genetic engineering will create two warring classes in society—the enhanced versus the naturals. Left-leaning bioethicists George Annas, Lori Andrews, and Rosario Isasi are brutally blunt about their fears of conflict between the genetically enhanced and unenhanced: "The new species, or 'posthuman,' will likely view the old 'normal' humans as inferior, even savages, and fit for slavery or slaughter. The normals, on the other hand, may see the posthumans as a threat and if they can, may engage in a preemptive strike by killing the posthumans before they themselves are killed or enslaved by them. It is ultimately this predictable potential for genocide that makes species-altering experiments potential weapons of mass destruction, and makes the unaccountable genetic engineer a potential bioterrorist."[27]

Another crowning achievement of the Enlightenment is the principle of tolerance, of putting up with people who look different, talk differently, worship differently, and live differently than we do. In the future, our descendants may not all be "natural" *Homo sapiens*. But they will still be moral beings who can be held accountable for their actions. Political liberalism is already the answer to bioconservative concerns about human and posthuman rights and interactions. In liberal societies the law is meant to apply equally to all, no matter how rich or poor, powerful or powerless, brilliant or stupid, enhanced or unenhanced. There is no reason to think that the same liberal political and moral principles that apply to diverse human beings today wouldn't apply to relations among future humans and posthumans.

But what if enhanced posthumans took the Nietzschean superman option? What if they really did see unenhanced people "as inferior, even savages, and fit for slavery or slaughter"? Let's face it, plenty of unenhanced humans have been quite capable of believing that millions of their fellow unenhanced humans were inferiors who needed to be eradicated. However, as liberal political institutions have spread around the world and strengthened, they have increasingly restrained technologically superior groups from automatically wiping out less advanced peoples (which was usual throughout most of history). I suspect that this felicitous dynamic will continue in the future as biotechnology, nanotechnology, and computational technologies progressively increase people's capabilities and widen their choices.

Proving that you can never make bioconservatives happy, they also worry about the obverse of genetic class warfare—what if *everybody* took advantage of genetic enhancement technologies? For example, when not propounding dystopian visions of genetically enhanced *Übermenschen* lording over poor naturals, McKibben is worried that genetic technologies will become cheap and widely available. Given the rapid pace of technological change, this is far more probable than genetic class warfare. *Therefore, safe affordable genetic medical technologies in an increasingly wealthy society are a recipe for eliminating genetic inequalities rather than perpetuating or exacerbating them.* Parents will have the option of giving their children the same genes for good health and smarter brains that other children get randomly now. Is this homogenization? Perhaps in some sense; but a world in which more people are smarter and healthier could hardly be an ethical or social disaster.

And if genetic class warfare or conformist homogenization are not enough, Fukuyama warns that humanity's Nietzschean will to power would tempt us to create subhuman slaves. Specifically, Fukuyama has suggested that biotechnology might be used to create half-human/half-chimpanzee slaves with the intelligence of a twelve-year-old boy.

Here Fukuyama is overlooking a few practical concerns, such as the fact that mothers willing to bear "subhuman slaves" in their wombs are likely to be scarce. And who would want a subhuman slave, anyway? Fully human slaves don't work out so well in the modern world. If you want real travel efficiency, you don't call for a slave-carried palanquin. You get into your Dodge Neon. If you need to write a letter, you don't summon your scribe. You fire up your Dell laptop.

Who doubts that ever-more-efficient and obedient machines will be cheaper and more practical solutions to the "servant problem" than any half-human/half-chimp slave would be? Besides, anyone who has ever tried to supervise the activities of a twelve-year-old for any length of time would likely pass on the opportunity to own one of Fukuyama's subhuman slaves.

Nevertheless, mixing human and animal genes and cells does pose some moral conundrums. It turns out that many genes are like animal-kingdom cassettes—they can be mixed and matched across species. A gene crucial to building a fruit fly's eye will also trigger eye development in a frog.[28] Now that both the human and mouse genomes have been sequenced, researchers know that 99 percent of mouse genes have homologues in humans; even more amazingly, 96 percent are present in the same order on both genomes.[29] Of course, those genes are expressed in very different ways, and mouse proteins, while similar, also differ in crucial ways.

First, consider the possibility of crossbreeding humans with other primates. There is some evidence that such mixing might succeed. As noted earlier researcher J. Michael Bedford reported in 1981 that human sperm could penetrate the protective outer membranes of gibbon eggs. So far, from what we know, no one has attempted to create a human/chimpanzee hybrid. But would that be wrong?

Bioethicist Joseph Fletcher once suggested that it would be ethical to create parahumans, that is, human/animal hybrids to do dangerous and demeaning jobs.[30] Is Fletcher's proposal all that different from training dolphins to find underwater explosives or using dogs to corral dangerous criminals?

Part of our unease arises from calling the creatures parahuman without defining which human characteristics might be added to them. Would giving an animal the ability to walk upright on two legs be morally problematic? Probably not. Would giving such creatures the ability to talk—that is, the capacity to understand and communicate with other language users—be morally problematic? That certainly raises the bar.

Human/animal crossbreeding is not the only way in which animals might be given the ability to talk. In 2002 researchers in Britain discovered that the FOXP2 gene in humans is required for articulate speech.[31] While it is not *the* language gene, it is certainly one of the genes necessary for the ability to talk. The proteins produced by the human FOXP2 gene differ by only two amino acids from the proteins produced by the FOXP2 gene in chimpanzees, gorillas, and orangutans. Would research that creates a transgenic chimpanzee with human FOXP2 genes elicit moral concern? Since FOXP2 orchestrates the actions of a variety of genes early in the development of the brains of human fetuses, and might have similar effects in chimpanzee fetuses, there may be grounds for ethical worries about such an experiment.

But what about just installing human brain cells directly into animals? Stanford University's Irving Weissman has injected human neural stem cells from aborted fetuses into the brains of fetal mice, where they have integrated and grown into human neurons and glia that intermingle with mouse brain cells, making up about 1 percent of the tissue in their brains. However, there is no evidence the chimeric mice began to contemplate the meaning of life. We need to give such chimeric mice no more or less moral consideration than we already give laboratory mice.

Weissman has said that he would like to inject human stem cells into the developing brains of fetal mice, with the goal of producing mouse brains composed chiefly of human brain cells.[32] Such mice might be useful for testing drugs to cure or prevent various human brain diseases. Since the brains would have the architecture of mouse brains, it is unlikely that they would become biotech Stuart Littles and exhibit any characteristics that would cause us moral concerns.

But what about injecting human brain stem cells into the developing brains of fetal chimpanzees? That's clearly a bit closer to the line, but if human cells are simply integrated into the typical architecture of a chimpanzee's brain, again, it would probably create no new ethical problems.

Besides the possibility of giving human characteristics to animals, injecting human stem cells into nonhumans could create other moral concerns. For example, stem cells might transdifferentiate into gamete-producing cells, and integrate themselves into the ovaries and testes of mice, where they would produce completely human eggs and sperm. One could imagine such chimeric male and female mice mating and producing a completely human embryo. Of course, that embryo would be unable to develop in the uterus of a mouse, so the world would not have to deal with the birth of a child whose mom was a rodent. But again, what if this research were done with larger chimeric animals—say, cows—that *could* possibly carry a human baby to term?

But chimeric mice could also be used to help people overcome infertility. Bone marrow stem cells from an infertile woman or man might be injected into a fetal mouse, where they could be transdifferentiated into gamete-producing cells. Gametes might be harvested from the mice and used in in vitro fertilization procedures to engender a child. Assuming it's medically safe, producing a child in this way would not be unethical.

Finally, it has to be asked: Would eating a liver composed chiefly of human liver cells grown in a sheep be cannibalism? I say yes; don't do it. Save them for transplants.

As humanity's biotechnological prowess increases, we will confront again and again the question of what, if any, limits should be placed on research that mixes human and animal genes, cells and tissues. The main ethical concern about such research is not the creation of improved and useful animals, but the risk of producing what would be, in effect, diminished human beings.

AN INHUMAN FUTURE?

It is almost impossible to parody the extent to which the antibiotech forces will go to scare us. And it isn't only our world they fear we are ruining—we are destroying the entire future of mankind. "*These are the most anti-choice technologies anyone's ever thought of,*" McKibben insists. "In widespread use, they will first rob parents of their liberty, and then strip freedom from every generation that follows. In the end, they will destroy forever the possibility of meaningful choice."[33] That claim is not only complete nonsense, it is exactly backward. McKibben's assertion echoes the argument made by British Christian apologist C. S. Lewis in his essay "The Abolition of Man." Lewis contended that one decisive generation that first masters genetic technologies would control the fate of all future generations.

In a sense this has always been true. Our ancestors—through their mating and breeding choices—determined for us the complement of genes that we all bear today. They just didn't know which specific genes they were picking. Fortunately, our descendants will have at their disposal ever-more-powerful technologies and the benefit of our own experiences—good and bad—to guide them in their future reproductive decisions. In no sense are they prisoners of our decisions now. Of course, there is one case in which future generations would be prisoners of our decisions now, and that's if we fearfully elect to deny them access to the benefits of biotechnology and safe genetic engineering.

Merely being obstructionists against human progress is bad enough, but some quarters of the intellectual Left go even farther, advocating neither genetic freedom not squelching biotech progress entirely, but plumping for top-down, government-driven genetic planning. This is tantamount to mandatory government subsidized eugenics in the name of equality. Yale University professor Ronald Dworkin is a supporter of such a troubling project.[34] And Dworkin is not alone. "Laissez-faire eugenics will emerge from the free choices of millions of parents," warns *Time* magazine columnist Robert Wright, who concludes, "The only way to avoid Huxleyesque social stratification may be for government to get into the eugenics business."[35]

A Brave New World of government eugenics is not an inevitable consequence of biomedical progress. We can choose whether we leave individuals free to make decisions about their biological futures or whether, in the name of equality or control, we give that power to centralized bureaucracies. Huxley's dystopia had no "laissez-faire eugenics" emerging from free choice; *Brave New World* is about a centrally planned society.

But McKibben (who, as we saw, admires authoritarianism if it helps curb technology) has worries far beyond those of simple political tyranny. In his mind, genetic engineering will enslave us on levels deeper than any despot ever could, turning our descendants into machines obeying the "ineradicable programs for our proteins" and finally, "we will live in a world where our humanity really has vanished."[36] McKibben seems stunningly ignorant of the fact that we *already obey* the ineradicable programs produced by our proteins; we just do not know what those programs are. Sadly, people learn all too often that what their present-day ineradicable protein programs are producing is disease, disability, and death.

But taking McKibben's concerns about our vanishing humanity seriously, if, for example, genetic engineering changes the range of typical human emotions, wouldn't that change what it means to be human? After all, no parent would want to give their children genes for emotional depression or a violent temper. University of Maryland's Institute for Philosophy and Public Policy senior research scholar Marc Sagoff claims that in order to qualify as being human, one must possess the emotional capacities that are characteristic of our species. Manipulating the human genome in a way that alters these capacities threatens the concept of human beings. If biotechnological manipulations removed our ability to feel anger, hate, or violence, "we would in an important sense not be human beings," argues Sagoff.[37] We'd still be members of the species *Homo sapiens* and we would still be moral beings, but not human beings as understood in previous ages.

But let's say that future genetic engineers discover that there is a gene that predisposes carriers to suicidal depression and that they can fix it. Would fixing it make subsequent generations nonhuman beings? After all, most people today do not fall into suicidal depressions, and those happy people are no less human than, say, the famously suicidal poet Sylvia Plath. Besides, depression can already be ameliorated for many people by means of Prozac or Paxil. Surely, taking serotonin reuptake inhibitors does not make people other or less than human. Sufferers of depression will tell you that the drugs do not make them feel less human; instead, they claim the drugs restore them to their true selves.

Most of us may already be incapable of berserker rage or religious ecstasy. Yet we are fully human beings, too. Sagoff cannot morally argue that parents must be forbidden to eliminate genes that predispose their offspring to suicidal depression on the grounds that that will deprive the rest of us of a sufficient number of great novelists in the future.

The corroded political root of opposition to biotech is the belief that things that are good for us as individuals should be forbidden in favor of the imagined good of the larger collective. McKibben, for example, openly eschews "hyperindividualism" in favor of one of the most destructive and oppressive political metaphors ever propounded. He wants us to think of "the human species as one large individual organism."[38] Never mind that in the last century ideologies founded on this organic principle of subordinating individual meaning to the good of the whole ended up killing tens of millions of people.

Of course, despite his affection for despotic technophobic regimes like Ming China and Tokugawa Japan, McKibben says he's for democracy as a way of choosing which limits to put on technological progress. "Happily for us, we have a system for dealing with competing ideas," he says. "It's called politics. We will have to *choose*."[39] But *we* don't have to choose; *each one* of us must be allowed to choose for him- or herself.

Chastened by the vast carnage of the religious wars in Europe, proponents of the Enlightenment project that spawned modern liberal democracies aimed to keep certain questions about the transcendent out of the public sphere in order to maintain public peace. Questions of the ultimate meaning and destiny of humanity were declared private concerns. However, bioconservatives such as Fukuyama and McKibben are calling that project into question. They argue that biotechnological choices should be subject to broad public scrutiny and stringent regulation. It is desperately important for concerns about biotechnology *not* to breach the Enlightenment understanding of what belongs in the private sphere and what belongs in the public. Technologies that touch on the big questions about birth, disease, death, and the meaning of life are exactly the kinds of activities that need protection from meddling by those who wish to force their visions of the transcendent on the rest of us.

Bioconservatives often accuse biotechnological progressives who would allow access to safe genetic enhancements of not being "worried about diminishing the sanctity of human life."[40] But who really has a higher regard for the sanctity of human life—those who, like Kass and McKibben, fatalistically counsel us to live with the often bum hands that nature deals us, or those who want to use genetic technologies to ameliorate the ills that have afflicted humanity since time immemorial? Respecting the sanctity of life doesn't require that we take whatever random horrors nature dishes out. Safe genetic engineering, when it becomes possible, strongly affirms the intrinsic value of human life by producing healthier, stronger, smarter people more equipped to enjoy their lives and to thrive during them.

No matter how advanced, technologies—including genetic enhancement—are not ends in themselves. They are means for individuals to build the best lives they can for themselves and their families. The decisions to use them or not are personal questions. These private arenas should not be open to public decision making. For a man who says he favors human freedom and choice, McKibben seems awfully eager to limit both.

WHOSE DECISION?

Let's bring this discussion back to the here and now by looking at what happened in the case of the Chicago mother we began with. Instead of applauding the mother for taking steps to insure that her child would be free of a terrible disease, the *Journal of the American Medical Association* published an editorial by a bioethicist and a physician who questioned whether the mother should bring a child into the world that she knows she will not be able to care for in a few years when she succumbs to Alzheimer's.

This is indeed a good question. But who is in a better position to decide it than the child's parents? Do these bioethical busybodies really think that this mother didn't agonize over this decision? After all, she is a genetic counselor, better informed than many parents of the risks and benefits of genetic testing and in vitro fertilization. Furthermore, she knows all about what it's like growing up with a single parent, since she lost her own father at an early age to the disease. And finally, her husband wanted this baby and is willing to raise her by himself when that time comes. In fact, thanks to PGD, they are now the parents of three children who will never succumb to the horrific disease that has devastated their mother's family.

The prospect of using PGD for any and all diagnosable diseases is particularly upsetting to some critics. "Today it's early-onset Alzheimer's. Tomorrow it could easily be intelligence, or a good piano player or many other things we might be able to identify the genetic factors for," complains Jeffrey Kahn, director of the University of Minnesota's Center for Bioethics. "The question is, whether we ought to."[41] In a *Washington Post* article, Kahn insists, "It's a social decision. This really speaks to the need for a larger policy discussion, and regulation or some kind of oversight of assisted reproduction."[42] But there is no reasonable ethical objection for using genetic testing to avoid disease. That's what medicine is supposed to do—cure and prevent disease.

Kahn is certainly wrong when he claims that the decision to use PGD by prospective parents is a "social decision" requiring more regulation. First of all, in the capacious sense implied by Kahn, any parent's decision to have a child even by conventional means has "social consequences" for us all. So would Kahn have neighbors, regulators, and bioethicists weigh in on everybody's reproductive decisions? Kahn would doubtless counter that unlike conventional reproduction, assisted reproduction involves the use of scarce medical resources that could be devoted to other purposes. But again, Kahn's notion of "social" could apply to anything. What if

Kahn disapproved of someone buying nonunion clothing or vacationing in the Caribbean rather than devoting his resources to building public parks or highways? In this case, the parents using assisted reproduction and PGD are spending their own money for the benefit of their own children to work with doctors who are freely exercising their skills. It's no more a "social decision" than if they had chosen to buy a Toyota Prius.

The usually more sensible University of Pennsylvania bioethicist Arthur Caplan asks, "Testing for diseases that are going to appear 30 or 40 years from now, does that make any sense, since people are mortal?"[43] Surely a mother whose family has been grievously afflicted with this disease and who faces it herself is in a better position to decide on that matter than even the most brilliant academic bioethicist. It is indisputably the height of moral rectitude for a parent to spare her children the terrible fate that this mother knows lies in store for herself. If it's all right to use efficacious medical treatments to cure a forty-year-old with Alzheimer's, it's all right to prevent him from getting it in the first place.

Caplan adds that PGD procedures are certain to prove highly controversial "because that's really getting into designing our descendants."[44] What horrors do such designer babies face? Longer, healthier, smarter, and perhaps even happier lives? It is hard to see any ethical problem with that. It is true for genetic engineering, as for all other technologies, that some people will misuse it; tragedies will occur. Given the sorry history of government-sponsored eugenics, it is vital that control over genetic engineering never be given to any governmental authority. But to use biotechnology and genetic engineering is not, by definition, to abuse it. This technology offers the prospect of ever-greater freedom for people, and should be welcomed by everyone who cares about human happiness and human flourishing.

Oxford University bioethicist Julian Savulescu is right when he reminds us, "The Nazis sought to interfere directly in people's reproductive decisions (by forcing them to be sterilized) to promote social ideals, particularly around racial superiority. Not offering selection for nondisease genes would indirectly interfere (by denying choice) to promote social ideals such as equality or 'population welfare.' There is no relevant difference between direct and indirect eugenics. The lesson we learned from eugenics is that society should be loath to interfere (directly and indirectly) in reproductive decisionmaking."[45]

People who benefit from the fruits of biotechnological progress in the future will be neither Frankenstein's monsters nor genetic robots. Rather, they will be our grateful descendants for whom we have eased the burdens of disease, disability, and early death—if only we choose not to slow or kill the development of this new technology. They will look back in wonder, and perhaps in horror, at those who would have denied them the blessings of biomedical progress.

CHAPTER 6

BIOTECH CORNUCOPIA

Improving Nature for Humanity's Benefit

Ten thousand people were killed and 10 to 15 million left homeless when a cyclone slammed into India's eastern coastal state of Orissa in October 1999. In the aftermath, the international aid organizations CARE and the Catholic Relief Society distributed a high-nutrition mixture of corn and soy meal provided by the US Agency for International Development to thousands of hungry storm victims.

Oddly, this humanitarian act elicited cries of outrage.

"We call on the government of India and the state government of Orissa to immediately withdraw the corn-soya blend from distribution," demanded Vandana Shiva, director of the New Delhi–based Research Foundation for Science, Technology, and Ecology. "The U.S. has been using the Orissa victims as guinea pigs for GM [genetically modified] products which have been rejected by consumers in the North, especially Europe."[1] Shiva's organization had sent a sample of the food to a US lab for testing to see if it contained any of the genetically improved corn and soybean varieties grown by tens of thousands of farmers in the United States. Not surprisingly, it did.

"Vandana Shiva would rather have her people in India starve than eat bioengineered food," responded C. S. Prakash, a professor of plant molecular genetics at Tuskegee University in Alabama.[2] Per Pinstrup-Andersen, then director general of the International Food Policy Research Institute, observed: "To accuse the U.S. of sending genetically modified food to Orissa in order to use the people there as guinea pigs is not only wrong; it is stupid. Worse than rhetoric, it's false. After all, the U.S. doesn't need to use Indians as guinea pigs, since millions of Americans have been eating genetically modified food for years now with no ill effects."[3]

Still, such ideological attacks on attempts to help the starving are common when it comes to GM foods. For another example, in the summer and fall of 2002, the African countries of Zimbabwe, Zambia, and Malawi were on the verge of famine. The United States offered to ship 540,000 tons of biotech corn to feed their 14 million people starving as a result of crop failures caused by a lethal combination of a prolonged drought and scandalously corrupt government policies. Calling the offered corn "poison," Zambian president Levy Mwanawasa at first refused the U.S. food aid for his 2.5 million countrymen facing starvation.[4] Eventually hungry Zambian villagers stormed the warehouses in which the corn was locked away and simply took it to feed their hungry families.[5]

Why would the leaders of these African nations risk starving millions of their citizens over fear of food that 290 million Americans have been eating safely since 1996? Because antibiotech activists such as Shiva and nongovernmental activist groups such as Greenpeace have been misleading the public about the alleged dangers of genetically improved crop varieties. And secondly, because the African leaders are worried that they might get caught in the crossfire of the growing trade war between the United States and the European Union (EU) over crop biotechnology. The EU essentially bans biotech crops and foods, so African leaders are being forced to choose between feeding their people or possibly losing European markets for their agricultural exports once the drought eases.

Shiva and her fellow activists also oppose golden rice, a crop that could help prevent blindness in half a million to 3 million poor children a year, and alleviate vitamin A deficiency in some 250 million people in the developing world. By inserting three genes, two from daffodils and one from a bacterium, scientists at the Swiss Federal Institute of Technology created a variety of rice that produces the nutrient beta-carotene, the precursor to vitamin A. In 2005 researchers boosted the beta-carotene content more than twentyfold. Agronomists at the International Rice Research Institute in the Philippines plan to crossbreed the variety, called golden rice because of the color produced by the beta-carotene, with well-adapted local varieties and distribute the resulting seeds to farmers all over the developing world. No one claims that golden rice is a panacea for correcting vitamin A deficiency, but it is a useful tool.

In June 2000, at a Capitol Hill seminar on biotechnology sponsored by the US Congressional Hunger Center, Shiva airily dismissed golden rice by claiming that "just in the state of Bengal 150 greens which are rich in vitamin A are eaten and grown by the women."[6] A visibly angry Martina McGloughlin, director of the biotechnology program at the University of California at Davis, retorted, "Dr. Shiva's response reminds me of . . . Marie Antoinette, [who] suggested the peasants eat cake if they didn't have access to bread."[7] Alexander Avery of the Hudson Institute's Center for Global Food Issues noted that nutritionists at UNICEF doubted it was physically possible to get enough vitamin A from the greens Shiva was recommending. Furthermore, it seems unlikely that poor women living in shanties in the heart of Calcutta could grow greens to feed their children.

Scientists trying to help the world's poor are appalled by the apparent willingness of biotechnology opponents to sacrifice people for their cause. At the annual meeting of the American Association for the Advancement of Science in February 2000, Ismail Serageldin, then director of the Consultative Group on International Agricultural Research, posed a challenge: "I ask opponents of biotechnology, do you want two to three million children a year to go blind and one million to die of vitamin A deficiency, just because you object to the way golden rice was created?"[8]

Shiva is one of the leaders in a growing global campaign against crop biotechnology, sometimes called green biotech (to distinguish it from medical biotechnology, known as red biotech). Antibiotech lobbying groups have proliferated faster than bacteria in an agar-filled petri dish: In addition to Shiva's organization, the Third World Network, and Greenpeace, they include the Union of Concerned Scientists, the Institute for Agriculture and Trade Policy, the Institute of Science in Society, the ETC Group (formerly the Rural Advancement Foundation International), the Ralph Nader–founded Public Citizen, the Council for Responsible Genetics, the Institute for Food and Development Policy, and that venerable fount of biotech misinformation, Jeremy Rifkin's Foundation on Economic Trends. But if their campaign against crop biotechnology is successful, its chief victims will be the downtrodden people on whose behalf the activists claim to speak.

Some antibiotech activists are not content to protest in the streets or pressure their legislatures. Gangs of antibiotech vandals with cute monikers such as Cropatistas and Seeds of Resistance have ripped up scores of research plots in Europe and the United States. The so-called Earth Liberation Front (ELF) burned down a crop biotech lab at Michigan State University on New Year's Eve in 1999, destroying years of work and causing $400,000 in property damage. Overall, the FBI estimates that ELF has perpetrated more than six hundred attacks and caused $43 million in damages since 1996.[9]

During the November 1999 World Trade Organization meeting in Seattle, at a standing-room-only "biosafety seminar" in the basement of a Seattle Methodist church, the British antibiotech activist Mae-Wan Ho declared, "This warfare against nature must end once and for all." Michael Fox, a vegetarian "bioethicist" from the Humane Society of the United States, sneered: "We are very clever little simians, aren't we? Manipulating the bases of life and thinking we're little gods. . . . The only acceptable application of genetic engineering is to develop a genetically engineered form of birth control for our own species."[10] This creepy declaration garnered rapturous applause from the assembled activists.

The global campaign against green biotech has had notable successes in recent years. Several leading food companies—including Gerber, Frito-Lay, and McDonald's—have been cowed into declaring that they will not use genetically improved crops to make their products. Since 1997 the EU has all but outlawed the growing and importing of biotech crops and food. In July 2003 the Cartegena Biosafety Protocol came into effect despite the refusal of the United States to ratify it. The protocol is an international treaty that mandates special labels for biotech foods and requires strict notification, documentation, and risk assessment procedures for biotech crops. Still, activists continue to push for a worldwide moratorium on planting genetically enhanced crops.

WHAT IS PLANT BIOTECHNOLOGY?

To decide whether or not the uproar over biotech is justified, you need to know a bit about how it works. In the last decade biologists and crop breeders have made enormous strides in their ability to select specific useful genes from various species and splice them into unrelated species. Previously, plant breeders were limited to introducing new genes through the time-consuming and inexact art of crossbreeding species that were fairly close relatives, for example, rye and wheat or plums and apricots. For each cross, thousands of unwanted and unknown genes would necessarily be introduced into a crop variety.

Years of backcrossing—breeding each new generation of hybrids with the original commercial variety over several generations—were needed to reduce the number of these unwanted genes so that the useful genes and characteristics chiefly remained. The new biotech methods are far more precise and efficient. Researchers can now take useful genes from any organism and insert them into other organisms. The plants they produce are variously described as "transgenic," "genetically modified," "genetically engineered," or "genetically enhanced."

Plant breeders using biotechnology have accomplished a great deal in only a few years. For example, they have created a class of highly successful insect-resistant crops by incorporating toxin genes from the soil bacterium *Bacillus thuringiensis* (*Bt.*). In nature, *Bt.* lurks on the leaves of plants, waiting to be ingested by foraging insects. Once it is eaten, it produces tiny crystals that puncture the insect's gut, killing it, and allowing the bacteria to feed on its carcass and multiply. These crystals are only toxic in the alkaline guts of insects, not in the acid stomachs of vertebrates. So *Bt.* is toxic largely to destructive caterpillars such as the European corn borer and the cotton bollworm; it is not harmful to birds, fish, mammals, or people. *Bt.* is so safe that for decades organic farmers have sprayed spores on crops as an insecticide. Now, thanks to some clever biotechnology, breeders have produced varieties of corn, cotton, and potatoes that make their own insecticide.

Another popular class of biotech crops incorporates an herbicide-resistance gene, a technology that has been especially useful in soybeans. Farmers can spray herbicide on fields holding such biotech crops and kill weeds without harming the crop plants. The most widely used herbicide is Monsanto's Roundup (glyphosate), which toxicologists regard as an environmentally benign chemical that degrades rapidly, only days after being applied. According to the United States Department of Agriculture (USDA) "glyphosate has extremely low toxicity to mammals, birds, and fish. The herbicides that glyphosate replaces are 3.4 to 16.8 times more toxic. . . . Thus, the substitution caused by the use of herbicide-tolerant soybeans results in glyphosate replacing other synthetic herbicides that are at least three times as toxic and that persist in the environment nearly twice as long."[11]

Even the radical Pesticide Action Network's Pesticides Database does not list Roundup as a PAN Bad Actor.[12] PAN Bad Actors are pesticides the group claims have been identified as carcinogens, reproductive or developmental toxicants, neurotoxic cholinesterase inhibitors, or groundwater contaminants. So if Roundup isn't a Bad Actor, that means PAN agrees that current research shows it isn't especially toxic, and doesn't cause cancer or reproductive problems. Another ecological advantage is that farmers who use Roundup Ready crops don't have to plow for weed control, which means they cause far less soil erosion.

Biotech crops are beloved by farmers—they are the most rapidly adopted new farming technology in history. The first generation of biotech crops was approved by the Environmental Protection Agency

(EPA), the FDA, and the USDA in 1995. The USDA estimates that in 2004 transgenic varieties accounted for 45 percent of corn acreage, 85 percent of soybean acreage, and 76 percent of cotton acreage in the United States.[13]

The International Service for the Acquisition of Agri-Biotech Applications (ISAAA) estimates that the global area planted in biotech crops in 2004 was 200 million acres (81 million hectares), up 20 percent from 2003. Today, some 8.25 million farmers in seventeen countries grow biotech crops. The ISAAA points out, "Notably, 90% of the beneficiary farmers were resource-poor farmers from developing countries, whose increased incomes from biotech crops contributed to the alleviation of poverty."

Nearly one-third of global biotech crop area was planted in developing countries. The total area planted in biotech crop varieties is up forty-seven-fold since 1996. The ISAAA projects that by 2010, 15 million farmers in thirty countries will adopt biotech crop varieties and that the acreage planted in biotech crops will rise to 150 million hectares (approximately 375 million acres).[14]

A study by the National Center for Food and Agricultural Policy estimated in 2002 that US farmers who have adopted biotech crop varieties produce an additional 4 billion pounds of food and fiber on the same acreage, improve farm income by $1.5 billion, and reduce pesticide volume by 46 million pounds. Evaluating forty case studies of twenty-seven crop varieties, the study further projected that farmers who adopt these new biotechnologically enhanced varieties could boost food production by an additional 10 billion pounds, improve farm income by another $1 billion, and reduce pesticide use by a total of 163 million pounds.[15]

With biotech soybeans, US farmers save an estimated $216 million annually in weed control costs and make 19 million fewer herbicide applications per year.[16] In addition, by using no-till farming made possible by herbicide-resistant biotech soybeans, farmers prevent 247 million tons of topsoil from eroding away.[17] It is estimated that herbicide-resistant biotech soybeans, canola, cotton, and corn varieties and insect-resistant biotech cotton reduced global pesticide use by 22.3 million kilograms of formulated product in 2000.[18] US cotton farmers alone avoided spraying 2.7 million pounds of insecticides and made 15 million fewer pesticide applications per year by switching to biotech varieties. Their net revenues increased by $99 million.[19] Researchers estimate that by preventing insect damage, *Bt.* corn increased yields by 66 million bushels in 1999.[20]

BIOTECH CROPS: SAFE AND SOUND

Despite irrational fears being peddled by activists, biotech crops are safe. Since being introduced in the mid-1990s, "there has not been a single adverse reaction to biotech food," said Lester Crawford, acting commissioner of the FDA, at a 2003 conference. "In the meantime, we've had tens of thousands of reactions to traditional foods."[21] In other words, to the US government's knowledge, no one has gotten so much as a sniffle, a stomachache, or a rash because of biotech foods.

Every independent scientific panel that has looked into the matter has concluded that foods made from biotech crops are safe to eat. Most Americans have been eating biotech crops since 1995. Today it is estimated that 70 percent of the foods on US grocery shelves are produced using ingredients from genetically enhanced crops. In April 2000 a National Research Council (NRC) panel issued a report that emphasized it could not find "any evidence suggesting that foods on the market today are unsafe to eat as a result of genetic modification."[22] North Carolina State University entomologist Fred Gould, who headed up that NRC panel, points out that the panel compared the safety of conventionally bred crops with genetically enhanced crops. "What we did find was that if you were comfortable with the level of safety in conventionally bred crops, we could say that there was no added concern from having genetically engineered crops with the same kind of traits," he says. "There's nothing special about the genetic engineering . . . process, that's going to make the plant less safe."[23] In other words, if you're afraid of biotech crops, you should be equally afraid of conventional crops.

As University of California, Davis biologist Martina McGloughlin remarked at a Congressional Hunger Center seminar in June 2000, the biotech foods "on our plates have been put through more thorough testing than conventional food ever has been subjected to."[24] According to a report issued by the House Subcommittee on Basic Research, "No product of conventional plant breeding . . . could meet the data requirements imposed on biotechnology products by U.S. regulatory agencies. . . . Yet, these foods are widely and properly regarded as safe and beneficial by plant developers, regulators, and consumers."[25] The report concluded that biotech crops are "at least as safe [as] and probably safer" than conventionally bred crops.[26] Even a review issued in the fall of 2001 of eighty-one separate European scientific studies of genetically modified organisms funded by the EU concluded that there was no evidence that genetically modified foods posed any new risks to human health or the environment.[27] Despite this finding, the EU nevertheless stringently restricts imports and the planting of biotech crops.

The Royal Society of London was one of seven scientific academies who together produced a report, *Transgenic Plants and World Agriculture*, that gave transgenic crops a clean bill of health. So has a report from the United Nations Food and Agriculture Organization (FAO). The Royal Society's vice president and biological secretary, Patrick Bateson, says that "it is disappointing to find a group like Greenpeace stating on its website that 'the risks are enormous and the consequences potentially catastrophic,' without offering any solid reasons to support such a claim."[28] Disappointing perhaps, but not too surprising since activist groups such as Greenpeace regularly use scare campaigns to drum up political and financial support.

BIOTECH FOOD FOR THE POOR

Today, as noted above, pest resistance and herbicide resistance, along with some disease resistance traits, are the chief improvements incorporated into biotech crops. And most of those enhancements have been made in leading commercial crops grown in developed countries, such as corn, soybeans, and cotton. The next frontier will be applying genetic enhancements to crops that feed the hungry in developing countries. However, this progress could be significantly slowed if a full-fledged trade war over biotech crops breaks out between the United States and Europe, increasing the risk of starvation for millions. The International Food Policy Research Institute (IFPRI) estimates that global food production must increase by 40 percent in the next twenty years to meet the goal of a better and more varied diet for a world population of some 8 billion people. As biologist Richard Flavell concluded in a 1999 report to the IFPRI, "It would be unethical to condemn future generations to hunger by refusing to develop and apply a technology that can build on what our forefathers provided and can help produce adequate food for a world with almost two billion more people by 2020."[29]

The FAO argues that crop biotechnology can be a "pro-poor agricultural technology," pointing out that crop biotechnology "can be used by small farmers as well as larger ones; it does not require large capital investments or costly external inputs and it is relatively simple to use. Biotechnologies that are embodied in a seed, such as transgenic insect resistance, are scale neutral and may be more affordable and easier to use than other crop technologies."[30] A 2004 Rand Corporation report agrees, noting that "the key component of the Gene Revolution technology is improved seed. This being the case, all farmers, small or large, should be able to take advantage of the Gene Revolution; theoretically, the Gene Revolution is scale-neutral, providing that one can pay for the seed."[31] Kenyan biologist Florence Wambugu agrees that crop biotechnology has great potential to increase agricultural productivity in Africa without demanding big changes in local practices.[32] A drought-tolerant seed will benefit farmers whether they live in Kansas or Kenya.

The world's poor farmers recognize this, even if the antibiotech activists who claim to speak for them don't. Thousands of poor Indian farmers nearly rioted in 2002 when the Indian government, spurred on by antibiotech activists, was poised to destroy the biotech pest-resistant cotton they had planted. Faced with this farmer revolt, the Indian government backed down.[33] The subsequent crops of biotech cotton performed spectacularly, boosting yields as much as 80 percent, reducing pesticide use by 70 percent, and increasing farmers' cotton-related income fivefold.[34] Now the Indian government has approved the biotech cotton.

Another way biotech crops can help poor farmers grow more food is by controlling parasitic weeds, an enormous problem in tropical countries. Cultivation cannot get rid of them, and farmers must abandon fields infested with them after only a few growing seasons. Herbicide-resistant crops, which would make it possible to kill the weeds without damaging the cultivated plants, would be a great boon to such farmers.

Immunizing crops against infectious diseases by incorporating genes for proteins from viruses and bacteria will also greatly benefit farmers both rich and poor. The papaya mosaic virus had wiped out papaya farmers in Hawaii, but a new biotech variety of papaya incorporating a protein from the virus is immune to the disease. As a result, Hawaiian papaya orchards are producing again, and the virus-resistant variety is being made available to developing countries. Similarly, scientists at the Donald Danforth Plant Science Center in St. Louis are at work on a cassava variety that is immune to the cassava mosaic virus, which killed half of Africa's cassava crop a few years ago. Biotech companies are granting broad licenses to use their patents to international and academic research institutes to enable the development of genetically enhanced crops such as cassava and rice, which are especially important to poor farmers in the developing world.[35]

Another recent advance with enormous potential is the development of biotech crops that can thrive in acidic soils, a large proportion of which are located in the tropics. Aluminum toxicity in acidic soils reduces crop productivity by as much as 80 percent.[36] Progress is even being made toward the Holy Grail of plant breeding: transferring the ability to fix nitrogen from legumes to grains. That achievement would greatly reduce the need for fertilizer. Biotech crops with genes for drought and salinity tolerance are also being developed. Already, researchers at Cornell University have identified techniques that could make plants more drought resistant.[37] Cornell researchers Ajay Garg and Ray Wu noted that South African "resurrection plants" produce the naturally occurring sugar trehalose, which helps them revive and resume growing after prolonged droughts. They added trehalose genes to rice, and the genetically improved rice withstood drought conditions much better than normal varieties do. Trehalose sugar is also found naturally in honey and mushrooms and is safe for people to eat.[38]

Further down the road, McGloughlin predicts,

> We will be able to use biotechnology to enhance nutritional content of crops such as protein, vitamins, minerals, and antioxidants, remove anti-nutrients, remove allergens, and remove toxins. We will also be able to enhance other characteristics such as growing seasons, stress tolerance, yields, geographic distribution, disease resistance, shelf life and other properties of production of crops. The ability to manipulate plant nutritional content heralds an exciting new area and has the potential to directly benefit developing countries.[39]

GROWING MEDICINE

Biotech crops hold promise not just for agriculture, but for medicine as well. Biologists at the Boyce Thompson Institute for Plant Research at Cornell University recently reported success in preliminary tests with biotech potatoes that would immunize people against diseases.[40] One protects against the Norwalk virus, which causes diarrhea, and another might protect against the hepatitis B virus, which afflicts 2 billion people. Plant-based vaccines would be especially useful for poor countries, which could manufacture and distribute medicines by having local farmers grow them. In order to maximize their effectiveness and standardize doses, some processing may be necessary—perhaps vaccines produced in crops could be canned and distributed without the need for expensive refrigeration.

Besides vaccines, crop biotechnology is also being developed as a way to produce therapeutic drugs cheaply. Research on pharmaceutical farming, or "pharming," is aimed at using plants to make human antibodies to treat cancer and arthritis. A number of companies are looking at ways to produce drugs in a variety of crops, including corn, rice, and tobacco. Once a plant has been genetically modified to make a particular drug, it would be relatively easy to scale up its manufacture simply by sowing more of it. One preliminary estimate suggests that producing 300 kilograms of therapeutic antibodies in plants would cost $10 million, compared to $80 million using conventional mammalian cell bioreactors.[41]

Since pharming aims to produce physiologically active medicines, segregating biopharm crops from food and feed crops is essential. Understandably, parents would be alarmed by a headline that declared that drugs had been found in little Jill's cereal. Consequently, the USDA has already set up guidelines for minimizing the chances that biopharm crops might get into the food supply. These include planting biopharm crops at different times than related food crops, establishing wide buffer zones so that biopharm and food crops don't cross-pollinate, and making sure that equipment used to process biopharm crops is dedicated to processing only such crops. Other proposals include genetically engineering the plants so that the drug is expressed only in stems and leaves and not seeds; engineering sterility into the biopharm crops so they can't cross-pollinate; and genetically modifying the plants to make them visually distinct from crop plants so they can be easily identified.[42] Nevertheless, it is unlikely that traces of biopharm crops slipping into the food supply would cause health or safety problems, since they would generally be diluted in the food-processing system to amounts well below those required to produce therapeutic effects.

Nevertheless, savvy biopharm companies will want to keep their valuable biopharm crops completely isolated. For example, a North Carolina company, Biolex, uses the tiny aquatic plant lemna, more commonly known as duckweed, to produce therapeutic compounds in a completely closed system. The duckweed, which resembles tiny clover plants without stems, grows in completely enclosed containers inside a factory and produces no pollen or seeds that can spread outdoors. Lemna is easily genetically engineered and can produce large quantities of therapeutic compounds because it doubles in numbers in about thirty-six hours. Biolex is producing the infection-fighting agent alpha interferon using duckweed, and has contracts with other pharmaceutical companies to produce monoclonal antibodies used to diagnose and treat diseases and a blood clot–busting protein.

BIOTECH ANIMALS

The benefits of genetic engineering aren't restricted to plants. Researchers are also looking at genetically engineering food animals and insect pests. The least controversial technology is the cloning of elite farm animals like premier milk cows or beef cattle to preserve their genes. For example, researchers at Texas A&M University cloned a twenty-one-year-old steer named Chance who was prized for his gentle nature. Naturally, his clone is known as Second Chance.[43] The National Academy of Sciences found no evidence that eating meat from cloned animals poses any health problems to people.[44] After all, clones are just younger twins of the animals whose meat and milk had already been safely consumed.

But animal biotechnology is not confined to just producing food. Canadian scientists have genetically modified goats to produce spider silk in their milk.[45] Spider silk is five times stronger than steel and could be used to manufacture products that require both light weight and exceptional strength. Cat-loving allergy sufferers like me are looking forward to the California biotech company ALLERCA's new nonallergenic felines, coming in 2008.[46] The cost for a sneeze-free tabby? About three thousand dollars. Inadvertent gene transfers are unlikely, since domestic animals don't generally interbreed with wild animals, though I suspect the value of ALLERCA's cats will decline as they spread their allergen-free genes through the neighborhood feline gene pools.

The biotech food animal closest to your plate is salmon engineered to grow faster and use 20 percent less feed. Developed by the Canadian company Aqua Bounty Technologies, their salmon produce growth hormone all year long instead of becoming dormant in the winter. This means they can grow to marketable size in twelve to eighteen months instead to two to three years. Fish today provide 15 percent of the animal protein eaten by the world's people. And fish are efficient at turning feed into protein. Farmed salmon already produce about one pound of salmon for every one and one-half pounds of feed consumed. It takes two pounds of feed to produce one pound of chicken and ten pounds of feed to produce one pound of beef.

Aqua Bounty's faster-growing salmon could help take pressure off the world's overstressed fisheries. Forty-seven percent of the oceans' wild fisheries are fully exploited, 18 percent are overexploited, and 10 percent are depleted, according the FAO's *State of the World Fisheries and Aquaculture 2002*.[47] Consequently, between 1996 and 2001, the world's capture fisheries' production essentially flattened to 94 million tons annually. Meanwhile, aquaculture production increased from 27 million tons in 1996 to 38 million tons in 2001. Intensifying aquaculture is the only way to keep up with the world's growing demand for fish while also protecting the world's wild fisheries.

Antibiotech activists fear that the fast-growing salmon could escape from offshore salmon pens and outcompete and interbreed with wild populations. Aqua Bounty, though, plans to sell and raise only sterile female fish, made thus by a process that forces fertilized salmon eggs to retain three sets of chromosomes. Such triploid fish cannot produce eggs. In addition, the faster-growing salmon might be raised in enclosed pens on land, eliminating the chance that they'd escape into the ocean. Aqua Bounty plans to apply to the FDA for approval for the fish in 2005. Developers of some thirty other genetically improved fish varieties could also soon seek regulatory approval.

While biotechnologists have successfully and safely modified crop plants to resist insects, others now plan to modify the insects themselves. Mosquitoes have been genetically modified so they can no longer harbor organisms that cause diseases in people, such as the malaria parasite, or viral diseases such as dengue fever and yellow fever. The tropical kissing bugs of Central and South America have been infected with genetically engineered bacteria that kill the Chagas trypanosome parasite carried by the bugs. The trypanosome that causes sleeping sickness carried by tsetse flies in Africa might be controlled in a similar fashion. Researchers at the University of California, Riverside are trying to stop an epidemic of Pierce's disease threatening California's vineyards. The disease bacterium is spread by a leafhopper pest called the glassy-winged sharpshooter. The Riverside scientists have modified a bacterium that lives in the guts of the sharpshooters so that they kill the bacteria that cause Pierce's disease.[48] Other researchers are trying to modify honeybees so they can resist the diseases and parasites that have devastated huge numbers of hives in the past decade.

This kind of genetic engineering is an extension of biological control strategies already in regular use. For example, pink bollworm moths that attack cotton and screwworm flies that infest livestock are controlled using sterile insect technique. Male moths and flies made sterile through irradiation are released in huge numbers to outcompete their wild rivals for mating with wild females, whose eggs then produce no progeny.[49] The US Forest Service controls gypsy moth infestations by spraying forests with a preparation of a natural virus that infects and kills only gypsy moths.[50]

Unlike crop plants, which can't typically compete with wild species, the goal of releasing genetically modified insect species will be that they successfully outcompete unmodified wild insects. Their potential ecological effects need to be very carefully considered before they are released. Potential negative effects will have to be balanced against the benefits expected from their release—which are considerable. For example, at least 300 million people contract malaria and perhaps as many as 3 million of them die from it every year. Having wild populations of malaria-carrying mosquitoes replaced with mosquitoes genetically modified to resist malaria would be a tremendous boon to humanity.

Releasing genetically modified insects and microorganisms to control diseases and pests will undoubtedly be modeled on successful programs like the biological control of the weed purple loosestrife. Biologists imported and released two leaf-feeding beetles and a root-eating weevil from Europe that eat only purple loosestrife. These insects were tested in laboratories to make sure that they would not endanger native North American plants before they were released. This effort at biological control has significantly reduced stands of the weed.[51] In a similar fashion, future genetically modified insects will be extensively tested and monitored in the lab before they are released.

Biotech opponents often warn portentously that genetically modified living organisms once released "cannot be recalled."[52] Of course, that's also true of unmodified organisms. While it is possible that genetically modified plants and animals could become disruptive when introduced into the wild, this risk must be evaluated in light of what we know about the history of introducing new species into ecosystems.

In the five hundred years since Columbus arrived in North America, some fifty thousand foreign species have become established here. These include nearly all our major crop plants—wheat, oats, soybeans, apples, oranges, and pears—and our livestock—cows, pigs, goats, sheep, and horses. Of course, some destructive pests have also found their way to our shores, but for the most part introduced species have not been very disruptive and have integrated well into our landscapes.[53] In fact, it is reasonable to expect that creatures such as genetically modified insects will usually be less disruptive than introduced exotic species, since their wild relatives will already be living in the ecosystem into which the modified animals are being introduced.

A recent study by researchers at Oregon State University noted, "Invasive exotic organisms represent the coordinated interaction and evolution of thousands of genes in a new environment, usually devoid of its pests and pathogen complex, [whereas] transgenic organisms result from one or a few intensively studied genes that encode highly specific traits."[54] If mosquitoes genetically modified to resist malaria or West Nile virus actually succeeded in replacing wild carriers, people would suffer just as many irritating bites. But they'd come down with fewer cases of illness.

BIOTECH FORESTS

Modern biotechnology has a lot to offer forestry as well. For example, the American chestnut was devastated by an introduced fungal disease that killed more than 3.5 billion trees in the first half of the twentieth century. These majestic trees, which could reach one hundred feet in height and five feet in diameter, had been the dominant hardwood species throughout the Appalachian Mountains. An enterprising squirrel, we are told, could travel from Maine to Georgia without touching the ground through the interlinked branches of chestnut trees. Now scientists at the University of Georgia and the State University of New York are investigating ways to insert blight-resistance genes from the Chinese chestnut into American chestnut artificial seed embryos. If successful, the American chestnut could be restored to the forests from which it has been missing for more than two generations.[55]

Other projects have genetically modified trees to produce less of the tough, stiff fiber lignin. This would make them better for making paper. Other trees have been modified to grow faster and straighter, to produce higher-quality lumber. Genetically improved trees growing faster on tree plantations would reduce the need to harvest wild trees and thus preserve natural forests. "The environmental benefit in a shift to planted from wild is that you could get all the wood the world needs pretty much from five, ten or twenty percent of the land used now," says Steve Strauss, professor of forest genetics and biotechnology at Oregon State University.[56] In other words, plantation forestry would enable humanity to leave 80 percent of forests wild.

Like other proposals to introduce genetically modified organisms into the environment, genetically modified trees have given rise to opposition, despite their many benefits for preserving and restoring woodlands. In 2001 the radical environmentalists of ELF destroyed genetically modified trees at the University of Washington at Seattle and a poplar farm in Oregon.[57] More moderate environmental activist groups such as the Sierra Club are calling for a worldwide moratorium on planting genetically modified trees.[58] They fear that the GM trees will somehow harm natural forests by interbreeding with their wild relatives. As in the case of biological control of pests, looking at current silvicultural practices can help make clear the benefits and risks involved with GM trees. "Many ecological criticisms of GM trees appear to be overstated," concludes a recent study by silviculturalists at Oregon State University. "The ecological issues expected from the use of GM poplars appear similar in scope to those managed routinely during conventional plantation culture, which includes the use of exotic and hybrid genotypes, short rotations, intensive weed control, fertilization and density control."[59] For example, choosing to plant a conventional poplar or a poplar genetically modified to produce less lignin will have far fewer ecological effects than choosing between planting a poplar, modified or not, and a conifer. "The specific changes in wood chemistry imparted by GM will be orders of magnitude less than the vast number of new chemicals that distinguish a pine from an aspen," notes the study.[60]

"TERMINATOR TECHNOLOGY"

Although the ecological risks of GM organisms are probably much fewer than those posed by novel unmodified organisms entering an ecosystem (and even those risks have turned out to be less serious than they are often portrayed as being), various bioconfinement techniques are being developed to limit the spread of GM organisms in the environment. A report from the National Academy of Sciences details some of them.[61] For example, GM crops and trees can be made sterile so they do not produce flowers, pollen, or seeds.

Given their concerns about the spread of transgenes, you might think biotech opponents would welcome innovations designed to keep them confined. Yet they became apoplectic when Delta Pine Land Company and the USDA announced the development of the Technology Protection System (TPS), a complex of three genes that makes seeds sterile by interfering with the development of plant embryos. TPS also gives biotech developers a way to protect their intellectual property: Since farmers couldn't save seeds for replanting, they would have to buy new seeds each year.

Because high-yielding hybrid seeds don't breed true, corn growers in the US and western Europe have been buying seed annually for decades. Thus TPS seeds wouldn't represent a big change in the way many American and European farmers do business. If farmers didn't want the advantages offered in the enhanced crops protected by TPS, they would be free to buy seeds without TPS. Researchers at the California biotech company Maxygen have proposed a kind of reverse genetic engineering that destroys all foreign genes in a crop when it is treated with a proprietary chemical. The system, dubbed "the Exorcist," consists of a protein called Cre that acts as "molecular scissors" and cuts out any DNA that lies between two particular DNA markers called IoxP. The idea is that crop biotechnologists would situate any gene they wanted to engineer into a crop—disease resistance, herbicide resistance, pest resistance—between a complex consisting of the genes for Cre and the IoxP markers. Also included would be a promoter sequence that turns on Cre when exposed to some specific agricultural chemical. So the farmer gets the benefit of pest resistance during the growing season, and should he later decide to sell his crop as GMO-free, he turns on the Cre protein, which destroys all the introduced genes including the Cre and IoxP genes. Furthermore, should the genetically enhanced crops have crossbred with conventional crops or weeds, the Exorcist complex would enable farmers to eliminate transgenes from them by treating them with the same agricultural chemical.[62] Similarly, seed companies could offer crops with transgenic traits that would be expressed only in the presence of chemical activators that farmers could buy if they thought they were worth the extra money. Ultimately, the market would decide whether these innovations were valuable.

If antibiotech activists are truly concerned about gene flow, they should welcome such technologies. If the pollen from crop plants incorporating TPS fertilized any neighboring conventional crops or weeds, any seeds produced would be sterile, so genes for traits such as herbicide resistance or drought tolerance couldn't be passed on.

This point escapes some biotech opponents. "The possibility that

[TPS] may spread to surrounding food crops or to the natural environment is a serious one," writes Vandana Shiva in her book *Stolen Harvest*. "The gradual spread of sterility in seeding plants would result in a global catastrophe that could eventually wipe out higher life forms, including humans, from the planet."[63] The biological ignorance of Shiva's claim is breathtaking. This dire scenario is not just implausible but biologically impossible: *TPS causes sterility; that means, by definition, that it can't spread.*

Despite the clear advantages TPS offers in preventing the gene flow that activists claim to be worried about, the Rural Advancement Foundation International (RAFI), now the ETC Group, quickly demonized it with the scare term "Terminator Technology." RAFI warned that "if the Terminator Technology is widely utilized, it will give the multinational seed and agrochemical industry an unprecedented and extremely dangerous capacity to control the world's food supply."[64] In 1998 farmers in the southern Indian state of Karnataka, urged on by Shiva and company, ripped up experimental plots of biotech crops owned by Monsanto in the mistaken belief that they were TPS plants. The protests prompted the Indian government to declare that it would not allow TPS crops into the country. That same year, twenty African countries declared their opposition to TPS at an FAO meeting. In the face of these protests, Monsanto, which had acquired the technology when it bought Delta Pine Land Company, declared that it would not develop TPS. Nevertheless, hope for eventually deploying TPS and other technologies that enable farmers to control the spread of transgenes is not dead. In February 2005 Canadian negotiators at an international conference on the UN's Convention on Biological Diversity initiated moves to lift the de facto moratorium and allow the testing and commercialization of TPS technology. However, Canada backed down in the face of opposition by most African countries, Austria, Switzerland, Peru, and the Philippines, who were spooked by activist assertions that such technologies would give seed corporations too much power over the world's food supplies.[65]

Meanwhile, researchers are developing another clever technique to prevent transgenes from getting into weeds and conventional crops through crossbreeding. Chloroplasts (the little factories in plant cells that use sunlight to produce energy) have their own small sets of genes. Researchers can introduce the desired genes into chloroplasts instead of into cell nuclei, where the majority of a plant's genes reside. The trick is that the pollen in most crop plants don't have chloroplasts; it is therefore very unlikely that a transgene confined to chloroplasts can be transferred through crossbreeding.[66]

One way to put the alleged health and environmental dangers posed by genetically engineering crops and other organisms in context is to consider the case of crops produced using radiation and chemical mutagenesis. In the first half of the twentieth century, hundreds of crop varieties popular with conventional and organic growers were created through mutations deliberately induced using radiation or chemicals. This method is obviously a far cruder and more imprecise way of creating new varieties, like using a sledgehammer instead of the delicate scalpel of biotechnology. The FAO lists twenty-three hundred officially released varieties of rice, wheat, barley, apples, citrus, sugarcane, and banana that have been produced using induced mutation to assist breeding.[67] Incidentally, in the United States the FDA doesn't review such crop varieties produced by radiation or chemicals for safety, yet there are no reports of even a single person having dropped dead from eating them. It is fair to conclude that if crops made by means of extremely crude mutations are safe—and all the evidence indicates they are—then crops produced using biotechnology are surely even safer.

BIOPANICS

Despite the wide agreement among scientific and medical organizations on the safety of biotech crops, activists still insist otherwise. They point for example, to a study by Arpad Pusztai, a researcher at Scotland's Rowett Research Institute, published in the British medical journal the *Lancet* in October 1999.[68] Pusztai found that rats fed one type of genetically modified potatoes (not a variety created for commercial use) developed immune system disorders and organ damage. The *Lancet*'s editors, who published the study even though two of six reviewers rejected it, apparently were anxious to avoid the charge that they were muzzling a prominent biotech critic. But the *Lancet* also published a thorough critique concluding that Pusztai's experiments "were incomplete, included too few animals per diet group, and lacked controls such as a standard rodent diet. . . . Therefore the results are difficult to interpret and do not allow the conclusion that the genetic modification of potatoes accounts for adverse effects in animals."[69] The Rowett Institute, which does mainly nutritional research, fired Pusztai for publicizing his results before they had been peer reviewed.

Activists are also fond of noting that the seed company Pioneer Hi-Bred produced a soybean variety that incorporated a gene—for a protein from Brazil nuts—that causes reactions in people who are allergic to nuts. The activists fail to mention that the soybean never got close to commercial release, because Pioneer Hi-Bred checked it for allergenicity as part of its regular safety testing and immediately dropped the variety.[70] The other side of the allergy coin is that biotech can remove allergens that naturally occur in foods such as nuts, wheat, potatoes, and tomatoes, making these foods safer.

In October 2000, activists seized on the news that a genetically modified corn variety called StarLink, approved by the EPA only for animal feed in the United States, had been found in two brands of taco shells, prompting recalls and front-page headlines. Lost in the furor was the fact that there was little reason to believe the corn was unsafe for human consumption—only an implausible, unsubstantiated fear that it might cause allergic reactions. After the fact, even Aventis, the company that produced StarLink, agreed that it was a serious mistake to have accepted the EPA's approval for animal use only. Because so many crops can be eaten by both people and livestock, most biotech proponents favor planting biotech feed crops only if they are determined safe for human consumption. In the end, the US Centers for Disease Control found that there was absolutely no evidence that anyone had suffered any adverse reaction to eating foods containing StarLink corn.[71]

In 2001 another biofood contamination scare story arose. Two activist researchers from the University of California, Berkeley claimed to have found genes from biotech corn in landrace varieties being grown by Mexican peasants. "DNA Contamination Feared," declared a headline in the *Washington Post*.[72] "Gene-altered DNA may be 'polluting' corn," warned *USA Today*.[73] "There is no scientific basis for believing that out-crossing from biotech crops could endanger maize biodiversity," says Luis Herrera-Estrella, a noted plant scientist and director of the Center for Research and Advanced Studies in Irapuato, Mexico. "Gene flow between commercial and native varieties is a natural process that has been occurring for many decades. Nor is there reason to believe that these genes will become fixed into landraces unless farmers select them for their increased productivity," adds Herrera-Estrella. "In the end, that would result in improving the native varieties."[74] The fact is that pollen flow between conventional varieties of crop plants and landrace varieties and wild relatives has been going on since the dawn of agriculture, yet no one has ever called such gene transfers "contamination." One or two traits carried by transgenes are unlikely to cause any more or less harm to landraces than the thousands of genes conventional varieties have no doubt been transmitting to them for decades.

In November 2004 a report issued by the North American Commission for Environmental Cooperation (CEC) under the North American Free Trade Agreement found no scientific basis for concern over transgenes "invading" Mexican corn landraces:

> There is no reason to expect that a transgene would have any greater or lesser effect on the genetic diversity of landraces or teosinte than other genes from similarly used modern cultivars. The scientific definition of genetic diversity is the sum of all of the variants of each gene in the gene pool of a given population, variety, or species. The maize gene pool represents tens of thousands of genes, many of which vary within and among populations. Transgenes are unlikely to displace more than a tiny fraction of the native gene pool, if any, because maize is an out-crossing plant with very high rates of genetic recombination. Instead, transgenes would be added to the dynamic mix of genes that are already present in landraces, including conventional genes from modern culti-

vars. Thus, the introgression of a few individual transgenes is unlikely to have any major biological effect on genetic diversity in maize landraces.[75]

Activists also cite environmental concerns as a reason to oppose plant biotechnology. Most notoriously, activists worry about how biotech corn pollen affects the Monarch butterfly. The global campaign against green biotech received a public relations windfall in 1999, when *Nature* published a study by Cornell University researcher John Losey that found that Monarch butterfly caterpillars died when force-fed milkweed leaves heavily dusted with pollen from *Bt.* corn.[76] Since then, at every antibiotech demonstration the public has been treated to flocks of activist women dressed fetchingly as Monarch butterflies. But when more realistic field studies were conducted, researchers found "there is no significant risk to monarch butterflies from environmental exposure to *Bt.* corn,"[77] and that the impact of *Bt.* corn pollen from current commercial hybrids on Monarch butterfly populations is negligible.[78] Corn pollen is heavy and doesn't spread very far, and milkweed grows in many places aside from the margins of cornfields. In the wild, Monarch caterpillars apparently know better than to eat corn pollen on milkweed leaves.

Furthermore, *Bt.* crops mean that farmers don't have to indiscriminately spray their fields with insecticides, which kill beneficial as well as harmful insects. In fact, studies show that *Bt.* cornfields harbor higher numbers of beneficial insects, such as lacewings and ladybugs, than do conventional cornfields. The fact is that pest-resistant crops kill mainly target species—that is, exactly those insects that insist on eating them.

Never mind; we will see Monarchs on parade for a long time to come. Meanwhile, a spooked EPA has changed its rules governing the planting of *Bt.* corn, requiring farmers to plant non–*Bt.* corn near the borders of their fields so that *Bt.* pollen doesn't fall on any milkweed growing there. But even the EPA firmly rejects activist claims about the alleged harms caused by *Bt.* crops. "Prior to registration of the first *Bt.* plant pesticides in 1995," it said in response to a Greenpeace lawsuit, "EPA evaluated studies of potential effects on a wide variety of non-target organisms that might be exposed to the *B.t.* toxin, e.g., birds, fish, honeybees, ladybugs, lacewings, and earthworms." The EPA concluded that these species were not harmed.[79]

The EPA's conclusion was reaffirmed in October 2003, when the British Royal Society's flagship scientific journal, the *Philosophical Transactions of the Royal Society*, reported the results of a three-year farm scale evaluation (FSE) study that compared conventional crops with genetically enhanced herbicide-resistant crops.[80]

Strangely, antibiotech activists actually claimed that the FSE results supported their demands for a total ban on genetically modified agriculture. "For years the GM corporations have been claiming that their crop would reduce weed killer use and benefit wildlife," Greenpeace executive director Stephen Tindale declared. "Now we know how wrong they were."[81] Tony Juniper, director of Friends of the Earth, argued, "These trials have shown that GM oilseed rape and beet cause more damage to the environment than even conventional crops. The maize results are at best inconclusive. Going ahead with the commercialization of any of these GM crops would be totally unacceptable."[82]

But do crops that are genetically enhanced to tolerate herbicides hurt the environment? Not really. The very limited question that FSE researchers were asked to investigate was whether or not there was any "difference between the management of GMHT [genetically modified herbicide tolerant] varieties and that of comparable conventional varieties in their effects on wildlife abundance and diversity."[83] To find out, farmers planted several score fields half with conventional varieties, and half with GMHT varieties. The researchers then looked at the abundance of weeds, invertebrates (insects, spiders, snails, etc.) and vertebrates (chiefly birds) living in the farm fields and along the uncultivated margins of the fields.

What did they find? That fields growing herbicide-tolerant beets and canola had fewer bees and butterflies. Why? Because bees and butterflies consume nectar, and the GMHT fields had fewer flowering weeds for them to feed on. The researchers noted that "the results for bees and butterflies relate to foraging preferences and might or might not translate into effects on population densities." In other words, bees and butterflies not surprisingly prefer to flit off to areas where flowers bloom and stay away from relatively weed-free fields. Meanwhile, another group of insects—springtails—increased in GMHT fields because they feed on dead plant matter, for example, the weeds killed by herbicides. Except for those groups, the researchers concluded, "The FSEs have shown that GMHT management has no strong effect on the majority of the higher taxa of aerial and epigeal arthropods."[84] Translation: Surface dwelling and flying invertebrates were largely unaffected by GM crops.

What about weeds? By engineering in herbicide tolerance, farmers can use safer, less toxic herbicides to control weeds throughout a crop's growing period. For conventional crops, farmers typically pretreat a field with herbicide to kill off weeds before or shortly after they plant. Since their crops are generally susceptible to herbicides, farmers are limited in the herbicides they can use once their conventional crop begins growing. So weeds that escaped the pretreatment continue to grow and compete with crop plants for nutrients and sunlight. Since herbicide-tolerant crops can be treated at any time, this means that farmers can more easily control weed infestations.

The Greenpeace claim that GM growing does not reduce the amount of weed killer used by farmers was shown by the FSE study to be dramatically false. Farmers used 48 percent less herbicide for GM beets, 43 percent less for maize, and not significantly less for canola, although in the United States GM canola farmers typically use 60 percent less herbicide than do conventional growers.

Biotech opponents try to sketch a dark portrait of herbicide-resistant crops, despite their obvious environmental benefits: a frightening future in which transgenic crops foster superpests—weeds bolstered by transgenes for herbicide resistance or pesticide-proof bugs that proliferate in response to crops with enhanced chemical defenses. As Martina McGloughlin notes, "The risk of gene flow is not specific to biotechnology. It applies equally well to herbicide resistant plants that have been developed through traditional breeding techniques."[85] Norman Ellstrand, a genetics professor at the University of California, Riverside notes that "there is now substantial evidence that at least 44 cultivated plants mate with one or more wild relatives somewhere in the world . . . crop-to-wild gene flow is not uncommon, and on occasion, it has caused problems. Would we expect transgenic plants to behave any differently? The answer is 'no.'"[86]

Well before the advent of genetically modified crops, herbicide resistance had been increasing over the decades in various weeds. As professor of plant physiology Jodie Holt, also from Riverside, observes, "As use of herbicides has increased, increased cases of selection for resistance in weeds have been documented. Since the first reported case of weed resistance in 1970, 258 weed species have evolved resistance to one or more of 18 herbicide classes."[87] Despite the fact that for nearly a decade tens of millions of acres have been sown with biotech crops, there have been precious few outbreaks of the much-dreaded "superweeds" caused by crossbreeding between biotech crops and wild plants. Even if an herbicide resistance gene did get into a weed species, most researchers agree that it would be unlikely to persist unless the weed were subjected to significant and continuing selection pressure—that is, sprayed regularly with a specific herbicide. And if a weed becomes resistant to one herbicide, it can be killed by another.

SCIENCE VERSUS VALUES

The FSE clearly provided some fascinating new information about the differences in conventional and genetically enhanced crops. The investigators should be applauded for rigorously and fairly answering the questions given to them. However, their findings do not ineluctably tell policymakers or the public what to do about genetically modified agriculture, no matter what the antibiotech activists may claim. The future direction of farming depends far more on value judgments and aesthetic concerns than it does on scientific studies. The true question is: What kind of landscapes do people prefer?

Consider that no matter what effects either conventional or GM crops have on wildlife, they pale in comparison to the impact that centuries of farming have had on farmland biology. In addition, the introduction of modern herbicides and pesticides fifty years ago made farmers' fields dramatically more productive, and comparatively weed and pest free.

Of course, the modern revolution in farming has boosted food production many fold, and makes food cheaper and more abundant than ever before in history. Few people would advocate doing away with conventional farming in order to boost wildlife populations, if by doing so we increased the risk of starvation. Farming, it's worth remembering, is the opposite of letting nature run wild—that's why agriculture is so much more productive than hunting and gathering.

Besides, the FSE researchers themselves point out that an alternative to banning biotech crops would be to manage the landscape to produce the sort of plants that support the preferred collection of insects, spiders, birds, and mammals. If protecting wildlife is your goal, the higher productivity of genetically enhanced crops means that less land has to be planted to grow food, thus leaving more land for nature. So there may be less wildlife in the fields farmers cultivate, but more across the whole landscape, after unused farmland has reverted to nature.

But why favor the wildlife that thrive in relatively open areas such as farms in the first place? Chopping down essentially all of Britain's forests to create farms had a far greater effect on wildlife than herbicides or genetically enhanced crops ever will. In the United Kingdom today, 85 percent of the total arable land surface is sown in crops. Why not get rid of farms entirely and restore Britain's once-dominant woodland species? After all, farmers in the EU have no business growing highly subsidized sugar beets, since the sugar they produce costs several times the world market price. It's the same with maize—farmers in Ohio can grow corn much more cheaply than the British can.

If antibiotech activists want to favor certain wildlife and not others, that is their choice. But they cannot make the case for their preferences by arguing that "science" has somehow proved their point.

THE DANGEROUS OPPOSITION TO BIOTECH FOOD

"Mutant people make mutant food," chanted antibiotech protesters as they paraded past the convention center in Boston where the members of the Biotechnology Industry Organization (BIO) held their annual meeting in March 2000. The thousands of protestors carried signs, making colorful suggestions such as

> "Monsanto stick your dick in something else. Stop fucking with us."
> "Clone this" emblazoned over an upraised middle finger.
> "Buck fio-tech."
> "Corporate greed spread by seed."
> "Biotechnology is giving pollution a life of its own."
> "No Gene Nazis."
> "Feed the Needy not the Greedy."

The protest parade also featured bushels of mutant tomatoes, a couple of green Tony the Tigers (cornflakes are made using biotech corn), various skeletal street puppets, and a giant papier-mâché carrot labeled "Hope." However, antibiotechnology protests are not just fun and games—they are life-and-death matters for some of the world's poorest people.

To date, the American public and policymakers have not generally succumbed to the scares and bogus concerns being peddled by antibiotech activists. Europe, however, is another matter entirely. A 2002 poll in the United Kingdom found that 51 percent of British consumers would avoid eating genetically enhanced foods, while 40 percent would not. However, 76 percent of respondents favored labeling biotech foods, while only 6 percent agreed with the US view that such foods need not be labeled.[88]

Since it is widely agreed by scientific experts around the world that food produced using biotech crops is safe, why are European regulators trying to ban it or stigmatize it using labels that the public would likely misconstrue as warning labels?

Current European resistance to genetically enhanced crops is generally traced to the concerns over food safety that erupted over the outbreak of mad cow disease in Britain and food contamination problems in Belgium in the 1990s. But there is a longer history to the EU's hostility to biotech. Starting in 1990, EU regulators, citing specious health concerns, fought against the importation of American beef and milk produced using biotech bovine growth hormone. After nearly a decade of dispute, the World Trade Organization (WTO) ruled in favor of the United States, allowing the United States to impose countervailing duties on more than $100 million in European food exports in retaliation.[89]

The EU justifies its ban of and import restrictions on biotech crops based on the "precautionary principle," which states that regulators do not need to show that a biotech crop is unsafe before banning it; they need only assert that it has not been proved harmless. The precautionary principle is best summed up as "regulate first, ask questions later."

The strictest interpretations of the precautionary principle jettison entirely the notion of tradeoffs, requiring that any new technology never cause any harm to the environment or human health. Of course, accurately predicting in advance the benefits and harms that a technology may one day produce is impossible. This inherent uncertainty means that opponents of a new technology can always stall its introduction by endlessly demanding more research be done to rule out even their most far-fetched fears.

Researchers Soren Holm and John Harris explain:

> As a principle of rational choice, the PP will leave us paralyzed. In the case of genetically modified (GM) plants, for example, the greatest uncertainty about their possible harmfulness existed before anybody had yet produced one. The PP would have instructed us not to proceed any further, and the data to show whether there are real risks would never have been produced. The same is true for every subsequent step in the process of producing GM plants. The PP will tell us not to proceed, because there is some threat of harm that cannot be conclusively ruled out, based on the evidence from the preceding step. The PP will block the development of any technology if there is the slightest theoretical possibility of harm. So it cannot be a valid rule for rational decisions.[90]

In other words, the only way to protect completely against unknown risks is never to do anything for the first time. While irrational in scientific terms, the precautionary principle is unfortunately all too rational in terms of satisfying the political needs of regulators and ideologically motivated activists.

The Cartegena Biosafety Protocol was drafted under the Convention on Biological Diversity and completed in 2000. The treaty, largely negotiated by environment ministers rather than trade ministers, oddly focuses almost entirely on international trade in living genetically modified organisms (LMOs). Specifically, that means trade in genetically enhanced crops and livestock. The protocol expressly incorporates the precautionary principle in its preamble and in articles 10 and 11 as justifications for signatories to limit the importation of LMOs such as grains and livestock.[91] The protocol requires that all shipments of biotech crops, including grains and fresh foods, carry a label saying they "may contain living modified organisms." This international labeling requirement is clearly intended to force the segregation of conventional and biotech crops.

Shortly after the protocol negotiations were completed in 2000, the European Commission issued a Communication from the Commission on the Precautionary Principle, which tried to explain how the EU would incorporate the principle in its regulatory systems.[92] Using the protocol and the commission's interpretation of the precautionary principle, the European Commission issued a set of new regulations in 2004 regarding biotech crops. Those regulations impose traceability and labeling requirements on all foods made using them.

Under these new regulations, all foods produced using ingredients derived from biotech crops and livestock, irrespective of whether they actually contain genetically modified DNA or proteins in the final product, must bear the following label: "This product contains genetically modified organisms."[93] Even corn syrup and soybean oil, which contain no detectable levels of DNA or biotech-derived proteins, will have to be labeled—in this case erroneously—as containing genetically modified organisms. Similar requirements are also proposed for feed grain that human beings will not eat.

Despite the fact that he admitted in October 2001 that "there is an irrational fear of GM food in the EU," European commissioner for health and consumer protection David Byrne justified these new regulations on consumer choice and protection grounds.[94] Indeed, even if no hazards from genetically improved crops have been demonstrated, don't consumers have a right to know what they're eating? This seductive appeal to consumer rights has been a very effective public relations gambit for antibiotech activists and European bureaucrats eager to expand their jurisdictions. If there's nothing wrong with biotech products, they ask, why shouldn't seed companies, farmers, and food manufacturers agree to label them?

The European biotech labeling rules set a new precedent in international trade treaties, which have never mandated labels merely to satisfy consumer curiosity. Government-required labels generally aim to alert consumers to health and safety concerns. Labels showing information on nutrition or allergens have always been based on objective, verifiable scientific evidence. Previously, companies have voluntarily labeled their products when they believe consumers want to know about some aspect of the manufacturing process, as with kosher or halal foods, "cruelty-free" products, and organic foods. Such voluntary process labeling alerts consumers who want this kind of information without imposing costs on those of us who don't. As with kosher, halal, cruelty-free, and organic products, the labeling of biotech foods has nothing to do with health or safety.

Although the EU claims its labels are not intended as warnings, they inevitably would feed ungrounded fears about the safety of biotech foods. "Even as we speak," said Tony Van der Haegen, a representative from the delegation of the European Commission to the United States, in a debate with me over biotech labeling, "the EU Agriculture Commission is assuring a delegation from Zambia that biotech foods are safe for them to eat." If so, I asked him, why isn't the EU Agriculture Commission also reassuring misinformed and frightened consumers in Europe that biotech foods are safe?[95]

At an International Policy Network conference in Geneva, Switzerland, Victor Bradley of the Canadian Department of Foreign Affairs and Trade flatly declared: "I have not run across any process labeling requirements [such as eco-labels or biotech labels] that had anything to do with consumers. They all have to do with establishing trade barriers."[96]

Activist scare tactics, such as coining ominous terms like "Frankenfoods," have created a climate in which many consumers would interpret labels on biotech products to mean that they were somehow more dangerous or less healthy than old-style foods. Biotech opponents hope labels would drive frightened consumers away from genetically modified foods and thus doom them. Then the activists could sit back and smugly declare that biotech products had failed the market test. Way back in November 1999, at the Seattle WTO meeting, an antibiotech protestor succinctly outlined the movement's strategy. "We're in a war," he said. "We're going to bury this first wave of biotech. The first battle is labeling. The second battle is banning it."[97]

In the United States, the biotech labeling campaign is a red herring anyway, because the USDA has issued some 554 pages of regulations outlining what qualifies as "organic" foods.[98] Among other things, the definition requires that organic foods not be produced using genetically modified crops. Thus, US consumers who want to avoid biotech products need only look for the "organic" label. Furthermore, there is no reason why conventional growers who believe they can sell more by avoiding genetically enhanced crops should not label their products accordingly, so long as they do not imply any health claims. The EU could adopt this approach as well, instead of imposing new regulations on genetically enhanced crops and foods.

In any case, labeling nonbiotech foods as such will not satisfy the activists whose goal is to force farmers, grain companies, and food manufacturers to segregate biotech crops from conventional crops. Such segregation would require a great deal of duplication in infrastructure, including separate grain silos, rail cars, ships, and production lines at factories and mills. It has been estimated that constructing the parallel infrastructure in order to comply with these regulations could cost the United States as much $4 billion. The StarLink corn problem is just a small taste of how costly and troublesome segregating conventional from biotech crops would be. In a study for the University of Guelph in Canada, KPMG Consulting estimated that a labeling mandate would add 35 to 41 percent to the prices of commodity grain, and raise the prices of processed foods by 9 percent to 10 percent—without any increase in safety.[99] Furthermore, mandatory crop segregation will lead to novel legal nightmares: If a soybean shipment is inadvertently "contaminated" with biotech soybeans, who is liable? If biotech corn pollen falls on an organic cornfield, can the organic farmer sue the biotech farmer?[100] One method of dealing with those issues is to set reasonable tolerances. Many activists and organic farmers advocate "zero tolerance" standards that in effect would outlaw genetically enhanced crops. Since every scientific body that has ever looked into the safety of biotech crops has found them safe, this would be an absurd requirement.

Crops have exchanged pollen for millennia, and they will continue to do so. Seed breeders have decades of experience in setting tolerances for seed purity. As Mark Condon, vice president for international marketing at the American Seed Trade Association, recently explained, "Seed purity certification standards were commonly set at ninety-eight percent to ninety-nine percent varietal purity levels or a standard of one percent to two percent adventitious genetic impurity."[101] It should be possible to maintain similar standards for organic crops. Since organic farmers set their own standards, they could easily adopt these tolerances and save themselves and conventional biotech farmers a lot of trouble.

BIOCONCERN OR TRADE BARRIERS?

The brewing US/EU trade war over biotech crops could imperil the whole WTO system of international trade, especially if socioeconomic considerations can be taken into account in the guise of implementing the precautionary principle. As the US/EU dispute over importing US beef produced using growth hormone indicates, the EU seems willing to accept the imposition of countervailing duties rather than comply with WTO rulings.

To maintain and expand a freer international trading system, we need a regulatory system based on scientific risk assessment. A "precautionary" approach will open the entire trading system to arbitrary interruptions, and capricious labeling requirements will proliferate. Such labels are unjustifiably stigmatizing and costly and offer no consumer health or safety benefits. Only objective scientific standards should be used, because regulations based on societal values can never be harmonized internationally.

As one tracks the activist campaign against green biotech, it becomes ever clearer that its leaders are not primarily concerned about safety. What they really dislike is free markets and globalization. "It is not inevitable that corporations will control our lives and rule the world," writes Shiva in *Stolen Harvest.*[102] In *Genetic Engineering: Dream or Nightmare?* Ho warns, "Genetic engineering biotechnology is an unprecedented intimate alliance between bad science and big business which will spell the end of humanity as we know it, and the world at large." The first nefarious step, according to Ho, will occur when the "food giants of the North" gain "control of the food supply of the South through exclusive rights to genetically engineered seeds."[103] In order to prevent this alleged takeover of the world's food supply, activists want to outlaw biotechnology patents.

"Patenting of life forms must be prohibited in order to preserve biodiversity, food security and indigenous peoples' rights and protect them from the corporate grip on genetic resources." declared a group of Green and Socialist parliamentarians at a press conference at the WTO meeting in Cancun, Mexico, in 2003.[104] Greenpeace is running a "No Patents on Life" campaign that appeals to inchoate notions about the sacredness of life. Knowing that no patents means no investment, biotech opponents declare that corporations should not be able to "own" genes, since they are created by nature.

Property rights over things such as land, houses, and cars are easily understood. Fences protect land and locks protect houses and cars from being stolen or misused by others. But intellectual property cannot be protected by fences and locks. Once an inventor has devised, say, a recipe for a powerful new drug, another drug manufacturer who finds out that recipe can easily make it. That means that the inventor, who spent the time, effort, and money to bring the benefit of a new cure to humanity, would not be compensated for his labor. Patents are designed to remedy that situation by providing strong incentives to inventors of beneficial products.

Patents are temporary monopolies, usually twenty years in duration, that aim to achieve two things. First, the patent system is much like peer-reviewed science; it's a system for disclosing to others how one achieved a certain result. A scientist does not get credit for a discovery until he publishes sufficient details so that other scientists can reproduce his experiments. Similarly, an inventor must disclose how to make the product, so that someone else can make it once the patent has expired. Second, by awarding a temporary monopoly to inventors, intellectual property rights encourage inventors to seek new discoveries by allowing them to make money either by licensing their patents for a fee to others or by giving them the exclusive right to make the product without competition for twenty years. Abraham Lincoln once described patents as "adding the fuel of interest to the fire of genius." Simply glancing up from this book to look around your house will show you how much we have benefited by this system; nearly every product that we use in everyday life was once patented.

Activists such as Andrew Kimbrell and Vandana Shiva assert that agricultural biotechnology companies are engaged in "biopiracy." They accuse transnational corporations such as Monsanto and Syngenta of stealing genes nurtured by the poor farmers of the world. The greedy corporations allegedly do this by patenting valuable genes found in local varieties of plants grown by traditional farmers. Then the companies try to sell the patented genes back to the poor farmers from which they took them. Sounds pretty unscrupulous, doesn't it?

What actually happens is that researchers at companies such as Monsanto screen a wide variety of plants seeking genes for things like disease resistance or the production of particular nutrients. Say, hypothetically, that the researchers find a gene in a local variety of rice in India that prevents a fungal disease endemic to India. Delighted, the corporate researchers have the technology to put the antifungus gene into a high-yielding but fungus-prone wheat variety. Farmers in India would have liked to grow the high-yield wheat, but didn't do so because of its susceptibility to fungus.

Genes are resources the same way that something like, say, copper is a resource. If I had a rock containing copper ore, perhaps I could use it as a paperweight. However, a copper rock is a much more valuable resource to someone who has the skill to mine, mill, refine, design, and market copper products, electrical wires, pots, and computer chips. Surely, it would be unreasonable for me to demand that the person who buys my copper rock and turns it into a pot give me the pot for free. The same goes for a beneficial gene, like the hypothetical antifungus gene that is inaccessible to Indian farmers because they have no way to get it from rice into wheat where it would be really helpful. Thanks to biotechnology, Indian farmers in my hypothetical example can now choose to grow (or not grow) the high-yield wheat without fear that their crop will be devastated by fungus. Seed companies liberate the useful genes and put them into high-yielding varieties that can boost poor farmers' productivity. Biopiracy is as much a fiction as copperpiracy.

Antiglobalization activists get it almost exactly backward. Intellectual property rights, far from being harmful to the poor, are in fact the foundation upon which many technologies that will help them rise from poverty to prosperity will be built. Amusingly, Shiva, while inveighing against "biopiracy," proudly claimed at a Congressional Hunger Center seminar that 160 varieties of kidney beans are grown in India.[105] Shiva is obviously unaware that farmers in India are themselves biopirates: Kidney beans were domesticated by the Aztecs and Incas in the Americas and brought to the Old World via the Spanish explorers.[106] In response to Shiva, C. S. Prakash pointed out that very few of the crops grown in India today are indigenous. "Wheat, peanuts, and apples and everything else—the chiles that the Indians are so proud of," he noted, "came from outside. I say, thank God for the biopirates." Prakash condemned Shiva's efforts to create "a xenophobic type of mentality within our culture" based on the fear that "everybody is stealing all of our genetic material."[107]

Shiva, Ho, and other activists rightly point to the inequities found in developing countries. They make the valid point that there is enough food today to provide an adequate diet for everyone if it were more equally distributed. They advocate land reform and microcredit to help poor farmers, improved infrastructure so farmers can get their crops to market, and an end to agricultural subsidies in rich countries that undercut the prices that poor farmers can demand.

Addressing these issues is important, but they are not arguments against green biotech. "The real issue is inequity in food distribution. Politics, culture, regional conflicts all contribute to the problem. Biotechnology isn't going to be a panacea for all the world's ills, but it can go a long way toward addressing the issues of inadequate nutrition and crop losses," said McGloughlin at the Congressional Hunger Center conference.[108]

Yet opponents of crop biotechnology can't stand the fact that it will help developed countries first. New technologies, whether reaping machines in the nineteenth century or computers today, are always adopted by the rich before they become available to the poor. The fastest way to get a new technology to poor people is to speed up the product cycle so the technology can spread quickly. Slowing it down only means the poor will have to wait longer. If biotech crops continue to catch on in developed countries, the techniques to make them will become available throughout the world, and more researchers and companies will offer crops that appeal to farmers in developing countries. In fact, the International Food Policy Research Institute released a report in January 2005 that found that research to develop genetically modified crops are in various stages of commercialization in fifteen developing countries on three continents. For example, researchers in China are working to produce genetically enhanced varieties of thirty different crops, Indian researchers are working on twenty-one different crops, and South Africans are modifying twenty different crops.[109]

Per Pinstrup-Andersen asked participants in the Congressional Hunger Center seminar to think about biotechnology from the perspective of people in developing countries:

We need to talk about the low-income farmer in West Africa who, on half an acre, maybe an acre of land, is trying to feed her five children in the face of recurrent droughts, recurrent insect attacks, recurrent plant diseases. For her, losing a crop may mean losing a child. Now, how can we sit here debating whether she should have access to a drought-tolerant crop variety? None of us at this table or in this room [has] the ethical right to force a particular technology upon anybody, but neither do we have the ethical right to block access to it. The poor farmer in West Africa doesn't have any time for philosophical arguments as to whether it should be organic farming or fertilizers or GM food. She is trying to feed her children. Let's help her by giving her access to all of the options. Let's make the choices available to the people who have to take the consequences.[110]

CHAPTER 7

CHANGING YOUR OWN MIND

The Neuroethics of Psychopharmacology

"We're on the verge of profound changes in our ability to manipulate the brain," says Paul Root Wolpe, a bioethicist at the University of Pennsylvania.[1] He isn't kidding. The dawning age of neuroscience promises not just new treatments for Alzheimer's and other brain diseases but enhancements to improve memory, boost intellectual acumen, and fine-tune our emotional responses. "The next two decades will be the golden age of neuroscience," declares Jonathan Moreno, a bioethicist at the University of Virginia. "We're on the threshold of the kind of rapid growth of information in neuroscience that was true of genetics 15 years ago."[2]

One man's golden age is another man's dystopia. One of the more vociferous critics of such research is Francis Fukuyama, who warns in his book *Our Posthuman Future* that "we are already in the midst of this revolution" and says that "*we should use the power of the state to regulate it*."[3] In a May 2002 cover story, the usually pro-technology *Economist* worried that "neuroscientists may soon be able to screen people's brains to assess their mental health, to distribute that information, possibly accidentally, to employers or insurers, and to 'fix' faulty personality traits with drugs or implants on demand."[4]

There are good reasons to consider the ethics of tinkering directly with the organ from which all ethical reflection arises. Most of those reasons boil down to the need to respect the rights of the people who would use the new technologies. Some of the field's moral issues are common to all biomedical research: how to design clinical trials ethically, how to ensure subjects' privacy, and so on. Others are peculiar to neurology. It's not clear, for example, whether people suffering from neurodegenerative disease can give informed consent to be treated or experimented on.

In May 2002 the Dana Foundation sponsored an entire conference at Stanford University on "neuroethics." Conferees deliberated over issues like the moral questions raised by new brain-scanning techniques, which some believe will lead to the creation of truly effective lie detectors. Participants noted that scanners might also be able to pinpoint brain abnormalities in those accused of breaking the law, thus changing our perceptions of guilt and innocence. Most nightmarishly, some worried that governments could one day use brain implants to monitor and perhaps even control citizens' behavior.

But most of the debate over neuroethics has not centered around patients' or citizens' autonomy, perhaps because so many of the field's critics themselves hope to restrict that autonomy in various ways. The issue that most vexes *them* is the possibility that neuroscience might enhance previously "normal" human brains.

The tidiest summation of their complaint comes from the conservative columnist William Safire. "Just as we have anti-depressants today to elevate mood," he wrote after the Dana conference, "tomorrow we can expect a kind of Botox for the brain to smooth out wrinkled temperaments, to turn shy people into extroverts, or to bestow a sense of humor on a born grouch. But what price will human nature pay for these nonhuman artifices?"[5]

Truly effective neuropharmaceuticals that improve moods and sharpen mental focus are already widely available and taken by millions. While there is some controversy about the effectiveness of Prozac, Paxil, and Zoloft, nearly 30 million Americans have taken them, with mostly positive results. In his famous 1993 book *Listening to Prozac*, psychiatrist Peter Kramer describes patients taking the drug as feeling "better than well." One Prozac user, called Tess, told him that when she isn't taking the medication, "I am not myself."[6]

Another increasingly popular neuropharmaceutical is modafinil, commercially available as Provigil in the United States. Modafinil is a psychostimulant drug that heightens alertness, brightens mood, and improves memory. Its side effects are minimal, and unlike amphetamines or cocaine, modafinil does not induce jitteriness, nor do users experience a cycle highs and lows. Modafinil also does not boost heart rates and blood pressure, and it appears to be nonaddictive. Clinically modafinil is used to treat narcolepsy, in which patients experience uncontrollable urges of daytime sleepiness. The US military is also very interested in modafinil because it can be safely used by troops and pilots who must stay alert for prolonged periods of time. It is also increasingly becoming a lifestyle drug used by professionals, shift workers, and students who want to remain alert for long periods of time. Researcher Barbara Sahakian from Cambridge University reported that in a double-blind trial of sixty healthy volunteers, modafinil significantly improved mental functioning, planning complex problems, recalling strings of numbers, and remembering abstract patterns.[7] "In my mind, it may be the first real smart drug," Sahakian asserts. "A lot of people will probably take modafinil. I suspect they do already."[8]

ONE PILL MAKES YOU SMARTER . . .

That's exactly what worries Fukuyama, who thinks that psychoactive drugs such as Prozac look a lot like *Brave New World*'s soma. The pharmaceutical industry, he declares, is producing drugs that "provide self-esteem in the bottle by elevating serotonin in the brain."[9] If you need a drug to be "your self," these critics ask, do you really have a self at all?

Another popular neuropharmaceutical is Ritalin, a drug widely prescribed to remedy attention-deficit/hyperactivity disorder (ADHD), which is characterized by agitated behavior and an inability to focus on tasks. Around 1.5 million schoolchildren take Ritalin, which recent research suggests boosts the activity of the neurotransmitter dopamine in the brain. Like all psychoactive drugs, it is not without controversy. Perennial psychiatric critic Peter Breggin argues that millions of children are being "drugged into more compliant or submissive state[s]" to satisfy the needs of harried parents and school officials.[10] For Fukuyama, Ritalin is prescribed to control rambunctious children because "parents and teachers . . . do not want to spend the time and energy necessary to discipline, divert, entertain, or train difficult children the old-fashioned way."[11]

Unlike the more radical Breggin, Fukuyama acknowledges that drugs such as Prozac and Ritalin have helped millions when other treatments have failed. Still, he worries about their larger social consequences. "There is a disconcerting symmetry between Prozac and Ritalin," he writes. "The former is prescribed heavily for depressed women lacking in self-esteem; it gives them more of the alpha-male feeling that comes with high serotonin levels. Ritalin, on the other hand, is prescribed largely for young boys who do not want to sit still in class because nature never designed them to behave that way. Together, the two sexes are gently nudged toward that androgynous median personality, self-satisfied and socially compliant, that is the current politically correct outcome in American society."[12]

Although there are legitimate questions here, they're related not to the chemicals themselves but to who makes the decision to use them. Even if Prozac and Ritalin can help millions of people, that doesn't mean schools should be able to force them on any student who is unruly or bored. But by the same token, even if you accept the most radical critique of the drug—that ADHD is not a real disorder to begin with—that doesn't mean Americans who exhibit the symptoms that add up to an ADHD diagnosis should not be allowed to alter their mental state chemically, if that's an outcome they want and a path to it they're willing to take.

Consider Nick Megibow, an undergraduate philosophy major at Gettysburg College. "Ritalin made my life a lot better," he reports. "Before I started taking Ritalin as a high school freshman, I was doing really badly in my classes. I had really bad grades, Cs and Ds mostly. By sophomore year, I started taking Ritalin, and it really worked amazingly. My grades improved dramatically to mostly As and Bs. It allows me to focus and get things done rather than take three times the amount of time that it should take to finish something."[13] If people like Megibow don't share Fukuyama's concerns about the wider social consequences of their medication, it's because they're more interested, quite reasonably, in feeling better and living a successful life.

What really worries critics like Safire and Fukuyama is that Prozac and Ritalin may be the neuropharmacological equivalent of bearskins and stone axes compared to the new drugs that are coming. Probably the most critical mental function to be enhanced is memory. And this, it turns out, is where the most promising work is being done. At Princeton, biologist Joe Tsien's laboratory famously created smart mice by genetically modifying them to produce more NMDA (N-methyl-D-asparate) brain receptors, which are critical for the formation and maintenance of memories. Tsien's mice were much faster learners than their unmodified counterparts. "By enhancing learning, that is, memory acquisition, animals seem to be able to solve problems faster," notes Tsien.[14] He believes his work has identified an important target that will lead other researchers to develop drugs that enhance memory. Later critics claimed that Tsien's genetically engineered mice besides having enhanced memories were also more sensitive to pain. Tsien has a different interpretation: he believes his "smart" mice are not more sensitive to pain but that they learn to avoid pain much more quickly than normal mice do.[15]

A number of biotech companies are already hard at work developing memory drugs. In December 2004, Helicon Therapeutics began Phase I clinical trials with its memory-enhancing compound code-named HT-0712.[16] The drug inhibits the activity of the PDE-4 enzyme that blocks the operation of another compound called cAMP response element binding protein (CREB), which plays a crucial role in consolidating long term memory.[17] Tim Tully, a professor of genetics at Cold Spring Harbor

Laboratory in New York who developed the drug, said, "If it proves safe and effective it could ultimately be used by people who want to learn a language or a musical instrument or even in schools."[18]

Helicon is not alone in pursuing memory enhancing drugs. Cortex Pharmaceuticals has developed a class of compounds called AMPA receptor modulators, which enhance the glutamate-based transmission between brain cells. Preliminary results indicate that the compounds do enhance memory and cognition in human beings. Memory Pharmaceuticals, cofounded by Nobel laureate Eric Kandel, is developing a calcium channel receptor modulator that increases the sensitivity of neurons and allows them to transmit information more speedily, and a nicotine receptor modulator that plays a role in synaptic plasticity. Both modulators apparently improve memory. Another company, Targacept, is working on the nicotinic receptors as well.

All these companies hope to cure the memory deficits that some 70 million baby boomers will suffer as they age. If these compounds can fix deficient memories, it is likely that they can enhance normal memories as well. Tsien points out that a century ago the encroaching senility of Alzheimer's disease might have been considered part of the "normal" progression of aging. "So it depends on how you define *normal*," he says. "Today we know that most people have less good memories after age 40, and I don't believe that's a normal process."[19]

EIGHT OBJECTIONS

And so we face the prospect of pills to improve our mood, our memory, our intelligence, and perhaps more. Bioethicists Paul Root Wolpe and Martha Farah note: "Enhancement techniques that affect brain function through more familiar and non-neuroscience-based interventions such as biofeedback, meditation, tutoring, or psychotherapy are not seen as objectionable, and, in fact, are often seen as laudable." So if it's clearly all right, and even laudable, to affect brain function in these ways, they ask, "What, then, are the objections to using pharmaceutical or other neurotechnological means to achieve the same ends as behavioral techniques?"[20]

Eight objections to such enhancements recur in neuroethicists' arguments. None of them is really convincing.

Neurological Enhancements Permanently Change the Brain

Erik Parens of the Hastings Center, a bioethics think tank, argues that it's better to enhance a child's performance by changing his environment than by changing his brain—that it's better to, say, reduce his class size than to give him Ritalin.[21] But this is a false dichotomy. Reducing class size is aimed at changing the child's biology, too, albeit indirectly. Teaching is supposed to induce biological changes in a child's brain, through a process called learning.

Fukuyama falls into this same error when he suggests that even if there is some biological basis for their condition, people with ADHD "clearly . . . can do things that would affect their final degree of attentiveness or hyperactivity. Training, character, determination, and environment more generally would all play important roles."[22] So can Ritalin, and much more expeditiously, too. "What is the difference between Ritalin and the Kaplan SAT review?" asks Dartmouth neuroscientist Michael Gazzaniga. "It's six of one and a half dozen of the other. If both can boost SAT scores by, say, 120 points, I think it's immaterial which way it's done."[23]

Neurological Enhancements Are Antiegalitarian

A perennial objection to new medical technologies is the one Parens calls "unfairness in the distribution of resources."[24] In other words, the rich and their children will have access to brain enhancements first, and will thus acquire more competitive advantages over the poor.

This objection rests on the same false dichotomy as the first. As Moreno puts it, "We don't stop people from giving their kids tennis lessons." If anything, the new enhancements might increase social equality. It's a lot easier for people to take pills than to pay for tutors and tennis lessons. Moreno also notes that neuropharmaceuticals are likely to be more equitably distributed than genetic enhancements, because "after all, a pill is easier to deliver than DNA."[25]

Neurological Enhancements Are Self-Defeating

Not content to argue that the distribution of brain enhancements won't be egalitarian enough, some critics turn around and argue that it will be *too* egalitarian. Parens has summarized this objection succinctly: "If everyone achieved the same relative advantage with a given enhancement, then ultimately no one's position would change; the 'enhancement' would have failed if its purpose was to increase competitive advantage."[26]

This is a flagrant example of the zero-sum approach that afflicts so much bioethical thought. Let's assume, for the sake of argument, that everyone in society will take a beneficial brain-enhancing drug. Their relative positions may not change, but the overall productivity and wealth of society would increase considerably, making everyone better off. Surely that is a social good.

Neurological Enhancements Are Difficult to Refuse

Why exactly would everyone in the country take the same drug? Because, the argument goes, competitive pressures in our go-go society will be so strong that a person will be forced to take a memory-enhancing drug just to keep up with everyone else. Even if the law protects freedom of choice, social pressures will draw us in.

For one thing, this misunderstands the nature of the technology. It's not simply a matter of popping a pill and suddenly zooming ahead. "I know a lot of smart people who don't amount to a row of beans," says Gazzaniga. "They're just happy underachieving, living life below their potential. So a pill that pumps up your intellectual processing power won't necessarily give you the drive and ambition to use it."[27]

Beyond that, it's not as though we don't all face competitive pressures anyway—to get into and graduate from good universities, to constantly upgrade skills, to buy better computers and more productive software, whatever. Some people choose to enhance themselves by getting a PhD in English; others are happy to stop their formal education after high school. It's not clear why a pill should be more irresistible than higher education, or why one should raise special ethical concerns while the other does not. Martha Farah from the University of Pennsylvania's Center for Cognitive Neuroscience is correct when she observes, "It would seem at least as much an infringement on personal freedom to restrict access to safe enhancements for the sake of avoiding the indirect coercion of individuals who do not want to partake."[28]

Neurological Enhancements Undermine Good Character

For some critics, the comparison to higher education suggests a different problem. We should strive for what we get, they suggest; taking a pill to enhance cognitive functioning is just too easy. As Fukuyama puts it: "The normal, and morally acceptable, way of overcoming low self-esteem was to struggle with oneself and with others, to work hard, to endure painful sacrifices, and finally to rise and be seen as having done so."[29]

"By denying access to brain-enhancing drugs, people like Fukuyama are advocating an exaggerated stoicism," counters Moreno. "I don't see the benefit or advantage of that kind of tough love."[30] Especially since there will still be many different ways to achieve things and many difficult challenges in life. Brain-enhancing drugs might ease some of our labors, but as Moreno notes, "there are still lots of hills to climb, and they are pretty steep."[31] Cars, computers, and washing machines have tremendously enhanced our ability to deal with formerly formidable tasks. That doesn't mean life's struggles have disappeared—just that we can now tackle the next ones.

Neurological Enhancements Undermine Personal Responsibility

Carol Freedman, a philosopher at Williams College, argues that what is at stake "is a conception of ourselves as responsible agents, not machines."[32] Fukuyama extends the point, claiming that "ordinary people" are eager to "medicalize as much of their behavior as possible and thereby reduce their responsibility for their own actions."[33] As an example, he suggests that people who claim to suffer from ADHD "want to absolve themselves of personal responsibility."[34]

But we are not debating people who might use an ADHD diagnosis as an excuse to behave irresponsibly. We are speaking of people who use Ritalin to *change* their behavior. Wouldn't it be more irresponsible of them to not take corrective action?

Neurological Enhancements Enforce Dubious Norms

There are those who assert that corrective action might be irresponsible after all, depending on just what it is that you're trying to correct. People might take neuropharmaceuticals, some warn, to conform to a harmful social conception of normality. Many bioethicists—Georgetown University's Margaret Little, for example—argue that we can already see this process in action among women who resort to expensive and painful cosmetic surgery to conform to a social ideal of feminine beauty.[35]

Never mind for the moment that beauty norms for both men and women have never been so diverse. Providing and choosing to avail oneself of that surgery makes one complicit in norms that are morally wrong, the critics argue. After all, people should be judged not by their physical appearance but by the content of their character.

That may be so, but why should someone suffer from society's slights if she can overcome them with a nip here and a tuck there? The norms may indeed be suspect, but the suffering is experienced by real people whose lives are consequently diminished. Little acknowledges this point, but argues that those who benefit from using a technology to conform have a moral obligation to fight against the suspect norm. Does this mean people should be given access to technologies they regard as beneficial only if they agree to sign on to a bioethical fatwa?

Of course, we should admire people who challenge norms they disagree with and live as they wish, but why should others be denied relief just because some bioethical commissars decree that society's misdirected values must change? Change may come, but real people should not be sacrificed to some restrictive bioethical utopia in the meantime. Similarly, we should no doubt value depressed people or people with bad memories just as highly as we do happy geniuses, but until that glad day comes people should be allowed to take advantage of technologies that improve their lives in the society in which they actually live.

Furthermore, it's far from clear that everyone will use these enhancements in the same ways. There are people who alter their bodies via cosmetic surgery to bring them closer to the norm, and there are people who alter their bodies via piercings and tattoos to make them more individually expressive. It doesn't take much imagination to think of unusual or unexpected ways that Americans might use mind-enhancing technologies. Indeed, the war on drugs is being waged, in part, against a small but significant minority of people who prefer to alter their consciousness in socially disapproved ways.

Neurological Enhancements Make Us Inauthentic

Parens and others worry that the users of brain-altering chemicals are less authentically themselves when they're on the drug. Some of them would reply that the exact opposite is the case. In *Listening to Prozac*, Kramer chronicles some dramatic transformations in the personalities and attitudes of his patients once they're on the drug. The aforementioned Tess tells him it was "as if I had been in a drugged state all those years and now I'm clearheaded."[36]

Again, the question takes a different shape when one considers the false dichotomy between biological and "nonbiological" enhancements. Consider a person who undergoes a religious conversion and emerges from the experience with a more upbeat and attractive personality. Is he no longer his "real" self? Must every religious convert be deprogrammed?

Even if there were such a thing as a "real" personality, why should you stick with it if you don't like it? If you're socially withdrawn and a pill can give you a more vivacious and outgoing manner, why not go with it? After all, you're choosing to take responsibility for being the "new" person the drug helps you to be.

AUTHENTICITY AND RESPONSIBILITY

"Is it a drug-induced personality or has the drug cleared away barriers to the real personality?" asks Wolpe.[37] Surely the person who is choosing to use the drug is in a better position to answer that question than some bioethical busybody.

This argument over authenticity lies at the heart of the neuroethicists' objections. If there is a single line that divides the supporters of neurological freedom from those who would restrict the new treatments, it is the debate over whether a natural state of human being exists and, if so, how appropriate it is to modify it. Wolpe makes the point that in one sense cognitive enhancement resembles its opposite, Alzheimer's disease. A person with Alzheimer's loses her personality. Similarly, an enhanced individual's personality may become unrecognizable to those who knew her before.

But are these kinds of changes so unusual? Already, many people experience a version of this process when they go away from their homes to college or the military. They return as changed people with new capacities, likes, dislikes, and social styles, and they often find that their families and friends no longer relate to them in the old ways. Their brains have been changed by those experiences, and they are not the same people they were before they went away. Change makes most people uncomfortable, probably never more so than when it happens to a loved one. Much of the neuro-Luddites' case rests on a belief in an unvarying, static personality, something that simply doesn't exist.

It isn't just personality that changes over time. Consciousness itself is far less static than we've previously assumed, a fact that raises contentious questions of free will and determinism. Neuroscientists are finding more and more of the underlying automatic processes operating in the brain, allowing us to take a sometimes disturbing look under our own hoods. "We're finding out that by the time we're conscious of doing something, the brain's already done it," explains Gazzaniga.[38] Consciousness, rather than being the director of our activities, seems instead to be a way for the brain to explain to itself why it did something.

Haunting the whole debate over neuroscientific research and neuroenhancements is the fear that neuroscience will undercut notions of responsibility and free will. Very preliminary research has suggested that many violent criminals do have altered brains. At the 2002 Stanford conference, *Science* editor Donald Kennedy suggested that once we know more about brains, our legal system will have to make adjustments in how we punish those who break the law. A murderer or rapist might one day plead innocence on the grounds that "my amygdala made me do it."[39] There is precedent for this: The legal system already mitigates criminal punishment when an offender can convince a jury he's so mentally ill that he cannot distinguish right from wrong.

Of course, there are other ways such discoveries might pan out in the legal system, with results less damaging to social order but still troubling for notions of personal autonomy. One possibility is that an offender's punishment might be reduced if he agrees to take a pill that corrects the brain defect he blames for his crime. We already hold people responsible when their drug use causes harm to others—most notably, with laws against drunk driving. Perhaps in the future we will hold people responsible if they fail to take drugs that would help prevent them from behaving in harmful ways. After all, which is more damaging to personal autonomy, a life confined to a jail cell or roaming free while taking a medication?

The philosopher Patricia Churchland examines these conundrums in her book *Brainwise: Studies in Neurophilosophy*. "Much of human social life depends on the expectation that agents have control over their actions and are responsible for their choices," she writes. "In daily life it is commonly assumed that it is sensible to punish and reward behavior so long as the person was in control and chose knowingly and intentionally."[40] And that's the way it should remain, even as we learn more about how our brains work and how they sometimes break down.

Churchland points out that neuroscientific research by scientists such as the University of Iowa's Antonio Damasio strongly shows that emotions are an essential component of viable practical reasoning about what a person should do. In other words, neuroscience is bolstering philosopher David Hume's insight that "reason is and ought only to be the slave of the passions." Patients whose affects are depressed or lacking due to brain injury are incapable of judging or evaluating between courses of action. Emotion is what prompts and guides our choices.

Churchland further argues that moral agents come to be morally and practically wise not through pure cognition but by developing moral beliefs and habits through life experiences. Our moral reflexes are honed through watching and hearing about which actions are rewarded and which are punished; we learn to be moral the same way we learn language. Consequently, Churchland concludes, "the default presumption that agents are responsible for their actions is empirically necessary to an agent's learning, both emotionally and cognitively, how to evaluate the consequences of certain events and the price of taking risks."[41]

It's always risky to try to derive an "ought" from an "is," but neuroscience seems to be implying that liberty—that is, letting people make choices and then suffer or enjoy the consequences—is essential for inculcating virtue and maintaining social cooperation. Far from undermining personal responsibility, neuroscience may end up strengthening it.

FOR NEUROLOGICAL LIBERTY

Fukuyama wants to "draw red lines" to distinguish between therapy and enhancement, "directing research toward the former while putting restrictions on the latter." He adds that "the original purpose of medicine is, after all, to heal the sick, not turn healthy people into gods."[42] He imagines a federal agency that would oversee neurological research, prohibiting anything that aims at enhancing our capacities beyond some notion of the human norm.

"For us to flourish as human beings, we have to live according to our nature, satisfying the deepest longings that we as natural beings have," Fukuyama told the Christian review *Books & Culture* in 2002. "For example, our nature gives us tremendous cognitive capabilities, capability for reason, capability to learn, to teach ourselves things, to change our opinions, and so forth. What follows from that? A way of life that permits such growth is better than a life in which this capacity is shriveled and stunted in various ways."[43] This is absolutely correct. The trouble is that Fukuyama has a shriveled, stunted vision of human nature, leading him and others to stand athwart the neuroscientific advances that will make it possible for more people to take fuller advantage of their reasoning and learning capabilities.

Like any technology, neurological enhancements can be abused, especially if they're doled out—or imposed—by an unchecked authority. But Fukuyama and other critics have not made a strong case for why *individuals*, in consultation with their doctors, should not be allowed to take advantage of new neuroscientific breakthroughs to enhance the functioning of their brains. And it is those individuals that the critics will have to convince if they seriously expect to restrict this research.

It's difficult to believe that they'll manage that. In the 1960s many states outlawed the birth control pill on the grounds that it would be too disruptive to society. Yet Americans, eager to take control of their reproductive lives, managed to roll back those laws, and no one believes that the pill could be outlawed again today.

Moreno thinks the same will be true of the neurological advances to come. "My hunch," he says, "is that in the United States, medications that enhance our performance are not going to be prohibited."[44] When you consider the sometimes despairing tone that Fukuyama and others like him adopt, it's hard not to conclude that on that much, at least, they agree.

Dartmouth's Gazzaniga is surely right when he concludes, "We can count on one universal: cognitive enhancement drugs will be developed . . . our society will absorb new memory drugs according to each individual's underlying philosophy and sense of self. Self-regulation of substances will occur; those few who desire altered states will find the drug, and those who don't want to alter their sense of who they are will ignore the availability of the drug. The government should stay out of it, letting our own ethical and moral sense guide us through the new enhancement landscape."[45]

CONCLUSION

THE AGE OF LIBERATION BIOLOGY

"The age of biotechnology is not so much about technology itself as it is about *human beings empowered by biotechnology*," emphasizes the President's Council on Bioethics in its report *Beyond Therapy*.[1] The report adds, "The dawning age of biotechnology" is providing humanity "greatly augmented power . . . not only for gaining better health but also for improving our natural capacities and pursuing our own happiness."[2] The council also acknowledges that the fondest hopes of biotechnology proponents are likely to be realized: "We have every reason to expect exponential increases in biotechnologies and, therefore, in their potential uses in all aspects of human life."[3]

Yet this prospect of biological liberation must be rejected, say influential bioconservatives on both the political Left and Right. They fear that the biotech revolution will undermine their devoutly held beliefs about the proper order of society and the goals of human life. So at the beginning of the twenty-first century we find ourselves in the remarkable position of having many of our leading intellectuals and policymakers arguing that their fellow citizens should be denied access to technologies they know will enable them and their families to live healthier, saner, and longer lives. These bioconservatives argue that a life span of seventy years is enough; that human freedom lies in randomly inheriting genes; that people should not be allowed to use assisted reproductive technologies to guarantee the health of their offspring; that deriving stem cells from five-day-old embryos is the moral equivalent of ripping the heart out of a thirty-year-old woman; that biotech crops are a danger to human health and the natural environment; and that drugs to improve memory and mood should be forbidden.

In order to slow and perhaps completely halt the development of the technologies they find threatening, bioconservatives urge the imposition of bans and moratoriums on research such as therapeutic cloning to produce stem cells. For example, the majority of the President's Council on Bioethics wants to impose such a moratorium because "the implications of proceeding or not proceeding are not clear." Consequently, "the proper attitude is modesty, caution, and moderation, expressed in a temporary ban to be revisited when time and democratic argumentation have clarified the matter."[4]

To control and perhaps even outlaw the technologies they fear, bioconservatives propose establishing "a national agency . . . with broad oversight, advisory, and decision-making authority" to regulate biomedical research. Francis Fukuyama puts it bluntly: "We need institutions with real enforcement power."[5] Fukuyama cites Britain's Human Fertilisation and Embryology Authority (HFEA) as the model for the power and scope of regulatory authority he and other bioconservatives want to establish.[6]

This is instructive. Consider the case of the Whitaker family from Sheffield, England, which perfectly illustrates the perils of allowing a government agency to interfere in a family's reproductive decisions. Michelle and Jayson Whitaker asked the HFEA for permission to use in vitro fertilization and preimplantation genetic diagnosis to produce a tissue-matched sibling for their son, Charlie, who suffers from a rare anemia and requires a blood transfusion every three weeks.[7] The HFEA refused, calling the procedure "unlawful and unethical" and ruling that tissue-matching is not a sufficient reason to attempt embryo selection.[8]

Desperate, the Whitakers came to the United States, where PGD is still legal. In June 2003 Michelle Whitaker gave birth to James, whose umbilical cord stem cells are immunologically compatible with Charlie's. The stem cells have now been transplanted, and according to his doctors, Charlie appears to be cured.[9] Please keep in mind that taking stem cells from James's umbilicus in no way endangered or harmed him.

In another case, the Mastertons of Scotland, who have four sons, lost their three-year-old daughter in a bonfire accident. "We lost a very precious member of our family and can never replace her. We don't see anything wrong or immoral in wanting science to help us have another little girl," says Alan Masterton.[10] However, the HFEA does. The agency refused to let them use sex selection as way to ensure the birth of a daughter.

In these cases, the HFEA's refusals were based not on issues of safety or efficacy but on the moral opinions of the members of the authority's governing panel. Such a regulatory authority necessarily turns differences over morality into win/lose propositions, with minority views overridden by the majority.

To the extent that new biotechnologies need regulation, agencies should be limited to deciding, as they have traditionally done, only questions about safety and efficacy. Regulatory agencies also certainly have an important role in protecting research subjects and patients from force and fraud by requiring that researchers obtain their informed consent. But when people of goodwill deeply disagree on moral issues that do not involve the prevention of force or fraud, it is not appropriate to submit their disagreement to a panel of political appointees.

The genius of a liberal society is that its citizens have wide scope to pursue their own visions of the good without excessive hindrance by their fellow citizens. In the United States we honor the free expression of moral diversity. Consequently, the federal government does not force Roman Catholic hospitals to provide abortions or contraception to their patients. Similarly, we recognize the right of adult Jehovah's Witnesses to refuse blood products and of Christian Scientists to refuse all medical treatments.

As history amply demonstrates, the public's immediate "yuck" reaction to new technologies is a very fallible and highly changeable guide to moral choices or biomedical policy. For example, in 1969 a Harris poll found that a majority of Americans believed that producing test-tube babies was "against God's will." However, less than a decade later, in 1978, more than half of Americans said that they would use in vitro fertilization if they were married and couldn't have a baby any other way.[11]

Another flaw in the idea of holding some sort of "national discourse" on biotechnological progress is that humans have terrible foresight. Evolution has so constituted us psychologically that we tend to imagine monsters lurking just over the horizon of technological progress, while failing to see that in reality the Promised Land lies close at hand. Therefore, if given the chance, "society" will often choose to block change. "Better the devil you know than the one you don't know," is all too often the adage preferred by many people.

Bioconservative intellectuals are fully cognizant of the tendency for our species to be suspicious of the new and the strange, and they clearly want to harness that suspicion as a strategy to restrain biotechnological progress. To that end, they advocate adopting the so-called precautionary principle. One canonical version reads: "When an activity raises threats of harm to human health or the environment, precautionary measures should be taken even if some cause and effect relationships are not fully established scientifically. In this context the proponent of an activity, rather than the public, should bear the burden of proof."[12] Adapted to the debate over biotechnology, it would mean that new biotechnological techniques such as stem cells, cloning, and sex selection should be presumed guilty until proven innocent, and those who propose a new intervention must bear the burden of showing that its promise outweighs its peril. And as bioethicist George Annas bluntly puts it, "The truth of the matter is that whoever has the burden of proof loses."[13]

The problem again is that humanity is terrible at anticipating benefits. Consider the optical laser. When the optical laser was invented in 1960, it was dismissed as "an invention looking for a job." No one could imagine what possible use this interesting phenomenon might be. Of course, now it is integral to the operation of hundreds of everyday products: it runs our printers and optical telephone networks, corrects myopia, removes tattoos, plays our CDs, opens clogged arteries, helps level our crop fields, etc. It's ubiquitous.

Or consider aspirin. Peter McNaughton, Sheild Professor of Pharmacology at Cambridge University, argues that the precautionary principle would have ruled out the introduction of aspirin. "This drug has considerable adverse side-effects, and would never be licensed today. The benefits, however, are enormous and growing—apart from the well-known treatment for inflammatory pain, there are uses in cancer, heart disease and prevention of deep vein thrombosis."[14]

As George Poste, director of the Biodesign Institute at Arizona State University explains:

> According to the precautionary principle, any activity in which a theoretical risk might exist should not be undertaken unless its outcome can be predicted fully in advance. The onus of proof rests with those who propose change. Since the full range of technical and social consequences of change can never be predicted with absolute precision, the principle is a prescription for paralysis.
>
> By institutionalizing caution it offers the allure of false security while precluding experimentation and change. In short, it creates categories of "forbidden knowledge."
>
> None of the remarkable advances in agriculture, aviation, chemistry, computing, energy, metallurgy and materials, medicine and telecommunications that have transformed living standards over the past two centuries would have overcome the hurdle of the precautionary principle.[15]

Clearly, his analysis applies to evaluating biomedical procedures as well. The greatest uncertainty about the possible harmfulness of organ transplants, Prozac, the birth control pill, in vitro fertilization, bypass surgery, X-rays, and almost everything else in the modern medical armamentarium was before they were first used. The same uncertainty clearly applies to new cancer therapies, rejuvenation by stem cells, and replacing mitochondria to prevent aging—in fact, the whole range of biotech advances discussed in this book.

The precautionary principle demands that we restrain technological developments not on the basis of what we know, but on the basis of what we do not know. Furthermore, the precautionary principle ignores the harms that come from not expeditiously proceeding with the development of new technologies. As noted earlier, the precautionary principle can be summed up as: Never do anything for the first time.

In fact, the way to make things clearer is to explore the advantages and disadvantages of the technology through research and cautious deployment. Asking citizens to decide an issue, even by means of "democratic argumentation," while they are still ignorant of its practicalities, benefits, and costs will not "clarify the matter"; only the additional knowledge gained from research can do that. As dissenters on the President's Council on Bioethics conclude, "Uncertainty over the potential of this research can only be overcome by doing the research."[16] Council member Janet D. Rowley also notes, "The effect of extending and expanding this moratorium [on therapeutic cloning to produce stem cells] will be to maintain our ignorance by preventing any research for four more years."[17]

Ultimately, the bioconservatives' call for democracy is disingenuous. It is at best a delaying tactic and at worst demagoguery—an appeal to the prejudices and emotions of an uninformed public. Besides, why should ethical decisions as personal as those involved in reproduction and medical care be decided by democratic vote?

History shows that the public's view of new technologies shifts as researchers and entrepreneurs make their benefits more widely known. As we've seen, had the public been allowed to vote on organ transplants or assisted reproduction at a stage of development comparable to where cloning and stem cell research stand today, they would have outlawed those biomedical advances, which now are widely approved. If researchers are correct about the benefits of stem cells and rejuvenation research, then the public will one day applaud those advances too. One cannot help but suspect that this is exactly what bioconservatives fear and want to forestall by calling for moratoriums they hope will become permanent.

In *Human Cloning and Human Dignity*, the council's bioconservative majority argues, "Our society needs more time to explore the full moral significance of taking such a step, to debate the moral and practical issues involved, and to seek a national consensus—about *all* research on early human embryonic (and fetal) life (not just that formed through cloning techniques)."[18] *All* research? Is this an effort to turn back the clock on such beneficial technologies as assisted reproduction or preimplantation genetic diagnosis of diseases in embryos? Does this debate include another fruitless and contentious effort to force a national consensus on the morality of abortion and contraception?

As Oxford University philosopher Nick Bostrom correctly argues,

> [T]he best way to avoid a Brave New World is by vigorously defending morphological and reproductive freedoms against any would-be world controllers. History has shown the dangers in letting governments curtail these freedoms. The last century's government-sponsored coercive eugenics programs, once favored by both the left and the right, have been thoroughly discredited. Because people are likely to differ profoundly in their attitudes towards human enhancement technologies, it is crucial that no one solution be imposed on everyone from above but that individuals get to consult their own consciences as to what is right for themselves and their families. Information, public debate, and education are the appropriate means by which to encourage others to make wise choices, not a global ban on a broad range of potentially beneficial medical and other enhancement options.[19]

The bioconservatives claim they want "wise public policy" to guide decisions about cloning and other biotechnologies. History has shown that truly wise public policy allows people, including biomedical researchers, maximum scope to pursue the good and the true in their own ways, in conformity with the dictates of their own consciences. The benefits of biotechnology are well known—the cure of diseases and disabilities for millions of sufferers; the production of more nutritious food with less damage to the natural environment; the enhancement of human physical and intellectual capacities—and all can be easily foreseen. It is the alleged dangers of biotechnology that are, in fact, vague, ill defined, and nebulous.

So what will the next fifty years of liberation biology hold? I foresee a convergence between science and art as the Promethean possibilities of genetic research become more widely available. The future will see miracles, cures, ecological restoration, vivid new art forms, and a greater understanding of the wellsprings of human compassion. But what can liberate can also tyrannize if misused. Evil minds may indeed try to pervert the gifts of bioscience by creating such monstrosities as designer plagues or a caste of genetic slaves. But history has shown that with vigilance—not with blanket prohibitions—humanity can secure the benefits of science for posterity while minimizing the tragic results of any possible abuse.

Nobody said the future would be risk free and the moral choices easy, but the future also brings wondrous new opportunities to cure disease, alleviate suffering, end hunger, and lengthen healthy lives. We would be less than human not to seize those opportunities.

NOTES

PREFACE

1. Leon Kass, "Why We Should Ban Human Cloning Now: Preventing a Brave New World," *New Republic*, May 21, 2001, http://www.tnr.com/052501/kass052101.html.

2. Farah Khan, "Genetics Seen to Breed Forms of Discrimination," Inter Press Service, September 4, 2001, http://www.commondreams.org/headlines01/0904-03.htm.

3. George W. Bush, "Remarks by the President on Stem Cell Research," August 9, 2001, http://www.whitehouse.gov/news/releases/2001/08/20010809-2.html.

4. Kass, "Why We Should Ban Human Cloning Now."

INTRODUCTION

1. Jeremy Rifkin, "Fusion Biopolitics," *Nation*, February 18, 2002, http://www.thenation.com/doc.mhtml?i=20020218&s=Rifkin.

2. William Kristol and Jeremy Rifkin, "Yes: Embryonic Clones Could Lessen the Value of Human Life in U.S.," *Detroit News*, July 21, 2002, http://www.detnews.com/2002/editorial.0207/22/a15-542125.htm.

3. Francis Fukuyama, *Our Posthuman Future: Consequences of the Biotechnology Revolution* (New York: Farrar, Straus & Giroux, 2002), p. 7.

4. Francis Collins, plenary lecture, American Association for the Advancement of Science Annual Conference, February 17, 2001; Larry Pressler, United Nations World Forum, remarks at panel discussion, "Advances in Biotechnology for the Human Genome and the Human Being: Technology, the Market, and Responsible Citizenry," September 8, 2000; Walter Link, United Nations World Forum, remarks at panel discussion, "Advances in Biotechnology for the Human Genome and the Human Being: Ethical, Philosophical, and Spiritual Issues," September 9, 2000.

5. Jeremy Rifkin, remarks at International Forum on Globalization Teach-In on Technology and Globalization, February 24, 2001.

6. Bill McKibben, *Enough: Staying Human in an Engineered Age* (New York: Times Books, 2003), p. 186.

7. Tom Athanasiou and Marcy Darnovsky, "The Genome as Commons," *WorldWatch* (July/August 2002), http://www.genetics-and-society.org/resources/cgs/200207_worldwatch_darnovsky.html.

8. Richard Hayes, "The Science and Politics of Genetically Modified Humans," *WorldWatch* (July/August 2002), http://www.genetics-and-society.org/resources/cgs/200207_worldwatch_hayes.html.

9. Leon Kass, "The Moral Meaning of Genetic Technology," *Commentary* (September 1999), http://www.commentarymagazine.com/Summaries/V108I2P34-1.htm.

10. Francis Fukuyama, "In Defense of Nature, Human and Non-Human," *New Perspectives Quarterly* (July 16, 2002), http://www.digitalnpq.org/global_services/global%20viewpoint/07-16-02.html.

11. Adam Wolfson, "Politics in a Brave New World," *Public Interest* (Winter 2001), http://www.thepublicinterest.com/archives/2001winter/article2.html.

12. Francis Fukuyama, "The World's Most Dangerous Ideas: Transhumanism," *Foreign Policy* (September/October 2004), http://georgeovermeire.nl/transhumanisme.nl/Fukayama.nl.

13. UNESCO, Universal Declaration on the Human Genome and Human Rights, November 11, 1997, http://portal.unesco.org/en/ev.php-URL_ID=13177&URL_DO=DO_TOPIC&URL_SECTION=201.html.

14. Nick Wadhams, "Divided U.N. Seeks Human Cloning Ban," *Washington Post*, March 8, 2005, http://washingtonpost.com/wp-dyn/articles/A17300-2005mar8.html.

15. Kass, "Moral Meaning of Genetic Technology."

16. Francis Fukuyama, "The End of History," *National Interest* 16 (Summer 1989): 18.

CHAPTER 1

1. Woody Allen, http:www.basicquotations.com/index.php?aid=136.

2. Leon Kass, "L'Chaim and Its Limits: Why Not Immortality?" *First Things* (May 2001), http://www.firstthings.com/ftissues/ft0105/articles/kass.html.

3. Ronald Bailey, "Intimations of Immortality," *Reason* (March 6, 1999), http://reason.com/opeds/rb030600.shtml.

4. Ronald Bailey, "Forever Young: The New Scientific Search for Immortality," *Reason* (August 2002), http://reason.com/0208/fe.rb.forever.shtml.

5. Ibid.

6. President's Council on Bioethics, *Beyond Therapy: Biotechnology and the Pursuit of Happiness* (Washington, DC: US Government Printing Office, 2003), p. 173.

7. Bailey, "Forever Young."

8. Bailey, "Intimations of Immortality."

9. Bailey, "Forever Young."

10. Bailey, "Intimations of Immortality."

11. Luis Rogers and Steve Farrar, "'Immortal' Genes Found by Science," *Times* (London), July 4, 1999, http://www.grg.org/rose.htm.

12. Nuno Arantes-Oliveira et al., "Regulation of Life-Span by Germ-Line Stem Cells in *Caenorhabditis elegans*," *Science* 295, no. 5554 (January 18, 2002), http://www.grg.org/UCSFWorms.htm.

13. Aubrey D. N. J. de Grey, "An Engineer's Approach to the Development of Real Anti-Aging Medicine," http://www.gen.cam.ac.uk/sens/manu16.pdf.

14. David Ewing Duncan, "The Pursuit of Longevity," *Acumen Journal of Sciences* 1, no. 2 (2003): 46.

15. De Grey, "An Engineer's Approach."

16. Jim Oeppen and James W. Vaupel, "Broken Limits to Life Expectancy," *Science* (May 10, 2002): 1030.

17. Robert A. Weale, "Biorepair Mechanisms and Longevity," *Journal of Gerontology: Biological Sciences* 59 (2004), http://biomed.gerontologyjournals.org/cgi/reprint/59/5/B449.

18. Kathryn Brown, "A Radical Proposal," *Scientific American Special Edition: The Science of Staying Young* 14 (2004): 31.

19. Caleb Finch and Sarah Crimmins, "Inflammatory Exposure and Historical Changes in Human Life-Spans," *Science* 305 (2004): 1736–39.

20. Ibid., p. 1737.

21. Christopher Wanjek, "Time in a Bottle: Anti-Aging Boosters Claim Their Products Can Turn Back the Clock. Independent Scientists Aren't Buying It," *Washington Post*, January 29, 2002, p. F01.

22. S. Jay Olshansky, Leonard Hayflick, and Bruce A. Carnes, "The Truth about Human Aging," *Scientific American* (May 13, 2002), http://www.sciam.com/article.cfm?chanID=sa004&articleID=0004F171-FE1E-1CDF-B4A8809EC588EEDF.

23. Ibid.

24. Roy L. Walford et al., "Physiologic Changes in Humans Subject to Severe, Selective Calorie Restriction for Two Years in Biosphere 2: Health, Aging, and Toxicological Perspectives," *Toxicological Sciences* 52, supp. (1999), http://www.cron-web.org/Walford_Biosphere1.pdf.

25. Maggie Fox, "Gene Tweaking Turns Couch Potato Mice into Racers," *USA Today*, August 27, 2004, http://www.usatoday.com/tech/science/genetics/2004-08-23-speedy-mice_X.htm.

26. Cynthia Kenyon, "Scientists Find What Type of Genes Affect Longevity," news release, University of California, San Francisco, June 29, 2003, http://pub.ucsf.edu/newsservices/releases/2003072131.

27. Leonard Guarente, "Forestalling the Great Beyond with the Help of SIR2," *Scientist* (April 26, 2004), http://www.the-scientist.com/2004/4/26/34/1.

28. "Eat What You Like, Stay Thin, Live to 120, No Cancer or Diabetes, May Be Possible Sooner Than We Think," *Medical News Today*, June 3, 2004, http://www.medicalnewstoday.com/medicalnews.php?newsid=9023.

29. David Secko, "'Longevity' Gene, Diet Linked," *Scientist* (June 18, 2004), http://www.biomedcentral.com/news/20040618/01; and "Who Says Chivalry Is Dead? Sir2 Fights against Aging in Mammals," Alzheimer Research Forum, June 17, 2004, http://www.alzforum.org/new/detail.asp?id=1031.

30. William J. Cromie, "Wine Molecule Slows Aging Process," *Harvard Gazette*, September 18, 2003, http://www.news.harvard.edu/gazette/2003/09.18/12-antiaging.html.

31. K. T. Howitz et al., "Small Molecule Activators of Sirtuins Extend *Saccharomyces cerevisiae* Lifespan," *Nature* 45, no. 6954 (September 11, 2003),

http://www.ncbi.nlm.nih.gov/entrez/query.fcgi?holding=npg&cmd=Retrieve&db=PubMed&list_uids=12939617&dopt=Abstract; and "Longevity: Of Red Wine, Diet, and a Long Life," *Science* STKE 200 (2003), http://stke.sciencemag.org/cgi/content/abstract/sigtrans;2003/200/tw356.

32. Alan D. Dangour, "Hormones and Supplements: Do They Work?" *Journal of Gerontology: Biological Sciences* 59 (2004), http://biomed.gerontologyjournals.org/cgi/content/full/59/7/B659.

33. Tory M. Hagen et al., "Feeding Acetyl-L-Carnitine and Lipoic Acid to Old Rats Significantly Improves Metabolic Function While Decreasing Oxidative Stress," *Proceedings of the National Academy of Sciences* 99, no. 4 (2002): 1870–75.

34. Gene Fowler and Bill Crawford, *Border Radio: Quacks, Yodelers, Pitchmen, Psychics, and Other Amazing Broadcasters on the American Airwaves* (New York: Limelight Editions, 1990), p. 22.

35. Steven Austad, *Why We Age: What Science Is Discovering about the Body's Journey through Life* (New York: John Wiley & Sons, 1997), p. 216.

36. Shaharyar Khan and Rafal Smigrodzki, "A Method of Delivery, Expression, and Replacement of DNA in Mitochondria," unpublished manuscript.

37. Aubrey D. N. J. de Grey, "Escape Velocity: Why the Prospect of Extreme Human Life Extension Matters Now," *PLoS Biology* 2, no. 6 (2004), http://www.plosbiology.org/plosonline/?request=get-document&doi=10.1371%2Fjournal.pbio.0020187.

38. Ibid.

39. Rita Colwell, remarks at National Science Foundation Small Wonders: Exploring the Vast Potential of Nanoscience Conference, March 19, 2002.

40. Ralph Merkle, "Nanotechnology and Medicine," in *Advances in Anti-Aging Medicine*, ed. Ronald M. Klatz (Larchmont, NY: Mary Ann Liebert, 1996), http://www.alcor.org/Library/html/NanotechnologyAndMedicine.html.

41. Robert A. Freitas, *Nanomedicine* (Austin, TX: Landes Bioscience, 1999).

42. Samuel Stupp, remarks at National Science Foundation Small Wonders: Exploring the Vast Potential of Nanoscience Conference, March 19, 2002.

43. Ben Best, "Vitrification in Cryonics," http://www.benbest.com/cryonics/vitrify.html.

44. Gregory Stock and Daniel Callahan, "Debates: Point-Counterpoint: Would Doubling the Human Life Span Be a Net Positive or a Negative for Us Either as Individuals or as a Society?" *Journal of Gerontology: Biological Sciences* 59 (2004), http://biomed.gerontologyjournals.org/cgi/content/full/59/6/B544.

45. Ibid.

46. William Nordhaus, "The Health of Nations: The Contribution of Improved Health to Living Standards," January 25, 2002, http://www.econ.yale.edu/~nordhaus/homepage/health_nber_1.doc.

47. Bill McKibben, *Enough: Staying Human in an Engineered Age* (New York: Times Books, 2003), p. 160.

48. Ibid., p. 159.

49. Brian Alexander, *Rapture: How Biotechnology Became the New Religion* (New York: Basic Books, 2003).

50. Kass, "L'Chaim and Its Limits."

51. National Institute on Aging, "Alzheimer's Disease Fact Sheet," February 2003, http://www.niapublications.org/pubs/pr01-02/adfact.html.

52. Gina Kolata, "Gains on Heart Disease Leave More Survivors, and Questions," *New York Times*, January 19, 2003, p. 1.

53. Daniel Callahan, "Visions of Eternity," *First Things* 133 (May 2003), http://www.firstthings.com/ftissues/ft0305/articles/callahan.html.

54. Bailey, "Forever Young."

55. Francis Fukuyama, "What Are the Possibilities and the Pitfalls in Aging Research in the Future?" SAGE Crossroads, February 12, 2003, http://www.sagecrossroads.net/021203transcript.cfm.

56. Ibid.

57. Leon Kass, "Ageless Bodies, Happy Souls: Biotechnology and the Pursuit of Perfection," *New Atlantis* 1 (Spring 2003), http://www.thenewatlantis.com/archive/1/kass/htm.

58. Leon Kass in *New York Times*, February 23, 1971, p. 17.

59. Ibid.

60. President's Council on Bioethics, *Beyond Therapy*, p. 187.

61. Ibid., p. 188.

62. Ibid., p. 189.

63. Ibid., p. 190.

64. Ibid., p. 192.

65. Ibid., p. 194.

66. Ibid., p. 195.

67. Ibid., p. 197.

68. Ibid., p. 192, emphasis mine.

69. President's Council on Bioethics, "Session 2: Beyond Therapy: Ageless

Bodies," transcript, March 6, 2003, http://www.bioethics.gov/transcripts/march03/session2.html.

70. Leon Kass, "The Case for Mortality," *American Scholar* 52, no. 2 (Winter 1977–78): 185.

71. Ibid., pp. 185, 187.

72. Ibid., pp. 184–85.

73. Bailey, "Intimations of Immortality."

74. Ibid.

75. Ibid.

76. Ibid.

77. President's Council on Bioethics, *Beyond Therapy*, pp. 192–93.

78. Bailey, "Forever Young."

79. Ibid.

80. Christine McGourty, "Eat Less to Live Longer and Healthier," *Sunday Times* (London), July 16, 2000.

CHAPTER 2

1. Mary Ann Schier, "Gene Therapy Patients Celebrate Five-Year Anniversary," *Conquest* 19, no. 1 (2004): 16, http://www.mdanderson.org/pdf/conquest_2004_summer.pdf.

2. Sophia Koliopoulous, "Gene Therapy: Medicine for Your Genes," The DNA Files, October 2001, http://www.dnafiles.org/about/pgm5/topic.html.

3. Marina Cavazzana-Calvo et al., "Gene Therapy of Human Severe Combined Immunodeficiency (SCID)–X1 Disease," *Science* 288 (April 28, 2000): 669–72.

4. Rick Weiss, "Boy's Cancer Prompts FDA to Halt Gene Therapy," *Washington Post*, March 4, 2005, p. A02.

5. Sheryl Gay Stolberg, "The Biotech Death of Jesse Gelsinger," *New York Times Magazine*, November 30, 1999, http://www.gene.ch/gentech/1999/Dec/msg00005.html.

6. Rick Weiss, "FDA Seeks to Penalize Gene Scientist," *Washington Post*, December 12, 2000, p. A14.

7. Sheila Connelly et al., "Sustained Phenotopic Correction of Murine Hemophilia A by In Vivo Gene Therapy," *Blood* 9, no. 91 (1998), http://www

.bloodjournal.org/cgi/content/full/91/9/3273?ck=nck; Jane D. Mount et al., "Sustained Phenotypic Correction of Hemophilia B Dogs with a Factor IX Null Mutation by Liver-Directed Gene Therapy," *Blood* 99, no. 8 (2002), http://www.bloodjournal.org/cgi/content/abstract/99/8/2670?view=abstract.

8. Anna Salleh, "Gene Therapy Restores Sight to Blind Dogs," ABC Science Online, April 30, 2001, http://www.abc.net.au/science/news/stories/s283715.htm.

9. R. Pawliuk et al., "Correction of Sickle Cell Disease in Transgenic Mouse Models by Gene Therapy," *Science* 294, no. 5550 (December 14, 2001), http://www.ncbi.nlm.nih.gov/entrez/query.fcgi?cmd=Retrieve&db=PubMed&list_uids=11743206&dopt=Citation.

10. Bing Wang, Juan Li, and Xiao Xiao, "Adeno-Associated Virus Vector Carrying Human Minidystrophin Genes Effectively Ameliorates Muscular Dystrophy in *mdx* Mouse Model," *Proceedings of the National Academy of Sciences* 97, no. 25 (December 5, 2000), http://www.pnas.org/cgi/content/abstract/97/25/13714?view=abstract.

11. S. C. Hyde et al., "Correction of the Ion Transport Defect in Cystic Fibrosis Transgenic Mice by Gene Therapy," *Nature* 362, no. 6714 (1993), http://www.ncbi.nlm.nih.gov/entrez/query.fcgi?cmd=Retrieve&db=PubMed&list_uids=7681548&dopt=Citation.

12. Catherine Manno et al., "AAV-Mediated Factor IX Gene Transfer to Skeletal Muscle in Patients with Severe Hemophilia B," *Blood* 101, no. 8 (2003), http://www.bloodjournal.org/cgi/content/abstract/101/8/2963?view=abstract.

13. "Copernicus Receives $1M from Cystic Fibrosis Foundation Therapeutics to Advance Development of a Non-Viral Gene Therapy for Cystic Fibrosis," *Business Wire*, January 20, 2005.

14. Michael Lasalandra, "Heart Patients May Ask FDA to Continue Gene Program," *Boston Herald*, July 27, 2001, p. 22.

15. Douglas W. Losordo et al., "Phase 1/2 Placebo-Controlled, Double-Blind, Dose-Escalating Trial of Myocardial Vascular Endothelial Growth Factor 2 Gene Transfer by Catheter Delivery in Patients with Chronic Myocardial Ischemia," *Circulation* 105 (2002), http://circ.ahajournals.org/cgi/content/abstract/105/17/2012?view=abstract.

16. Antonio Musaro et al., "Localized Igf-1 Transgene Expression Sustains Hypertrophy and Regeneration in Senescent Skeletal Muscle," *Nature Genetics* 27, no. 2 (2001), http://www.nature.com/doifinder/10.1038/84839.

17. Elisabeth R. Barton Davis et al., "Viral Medicated Expression of Insulin-

like Growth Factor I Blocks the Aging-Related Loss of Skeletal Muscle Function," *Proceedings of the National Academy of Science* 95, no. 26 (1998), http://www.pnas.org/cgi/content/full/95/26/15603.

18. Richard A. Lovett, "Gene Doping Is Shaping Up as the Next Big Issue in Debate over Performance Enhancers," *San Diego Union-Tribune*, July 21, 2004, http://www.signonsandiego.com/news/science/20040721-9999-lz1c21doping.html.

19. Alan Zarembo, "DNA May Soon Be in Play," *Los Angeles Times*, August 27, 2004, p. A1.

20. Ibid.

21. Christie Aschwanden, "Gene Cheats," *New Scientist* (January 15, 2000): 24.

22. Rob Stein, "Muscle-Bound Boy Offers Hope for Humans: Scientists Work to Isolate Secrets of a Genetic Mutation That Could Alleviate Weakness Accompanying Disease and Aging," *Washington Post*, June 28, 2004, p. A07.

23. Yong-Xu Wang et al., "Regulation of Muscle Fiber Type and Running Indurance by PPARδ," *PLoS Biology* 2, no. 10 (2004), http://www.plosbiology.org/plosonline/?request=get-document&doi=10.1371%2Fjournal.pbio.0020294.

24. Philip Cohen, "Marathon Mice Can Run and Run," *New Scientist* (August 28, 2004): 12.

25. Alison Korn, "Genetic Test Can Guide Young Athletes," *Calgary Herald*, January 30, 2005, p. A16.

26. Sylvia Pagán Westphal, "Just Add a Chromosome…," *New Scientist* (June 19, 2004): 10.

27. Judith Campisi, personal communication, June 2004.

28. Sylvia Pagán Westphal, "Cancer Gene Therapy Is First to Be Approved," *New Scientist* (November 28, 2003), http://www.newscientist.com/article.ns?id=dn4420.

29. Adam Luck, "Chinese Gene Therapy Offers Hope to Terminally Ill Cancer Patients," *Telegraph* (London), July 4, 2004, http://www.health.telegraph.co.uk/news/main.jhtml?xml=/news/2004/07/04/wgene04.xml.

30. "The Genesis of Gendicine: The Story behind the First Gene Therapy," *Bio-Pharm International* (May 2004), http://www.nbsc.com/files/papers/BP5-08-04Ae.pdf.

31. Renuka Rayasam, "Introgen's Drug Enters FDA Phase: Austin Company's Gene Cancer Therapy Could Be 1st Approved in US," *Austin American-Statesman*, December 24, 2004, p. A1.

32. "Advexin Vaccine Plus Chemotherapy Provides Impressive Responses in Locally Advanced Breast Cancer," *Breast Cancer News*, December 2004, http://professional.cancerconsultants.com/oncology_breast_cancer_news.aspx?id=32797.

33. Stephen J. Swisher et al., "Induction of p53-Regulated Genes and Tumor Regression in Lung Cancer Patients after Intratumoral Delivery of Adenoviral p53 (INGN 201) and Radiation Therapy," *Clinical Cancer Research* 9 (January 2003), http://clincancerres.aacrjournals.org/cgi/content/abstract/9/1/93.

34. Thomas Okarma, personal communication, June 2004.

35. Wei Lu et al., "Therapeutic Dendritic-Cell Vaccine for Chronic HIV-1 Infection," *Nature Medicine* 10 (November 28, 2004), http://www.nature.com/cgi-taf/DynaPage.taf?file=/nm/journal/v10/n12/abs/nm1147.html.

36. Okarma, personal communication, June 2004.

37. Campisi, personal communication, June 2004.

38. Donald Kennedy, "Breakthrough of the Year," *Science* 298, no. 5602 (December 20, 2002): 2283; Thomas Tuschl, "10 Emerging Technologies That Will Change Your World," *Technology Review* (February 2004), http://www.technologyreview.com/articles/04/02/emerging0204.asp?p=8.

39. Kimberly M. Keene et al., "RNA Interference Acts as a Natural Antiviral Response to O'nyong-nyong Virus (*Alphavirus*; Togaviridae) Infection of *Anopheles gambiae*," *Proceedings of the National Academy of Science* 101, no. 49 (December 7, 2004), http://www.pnas.org/cgi/content/abstract/101/49/17240.

40. Safia Wasi, "RNA Interference: The Next Genetics Revolution?" *Horizon Symposia*, May 2003, http://www.nature.com/horizon/rna/background/pdf/interference.pdf.

41. National Institute of Neurological Disorders and Stroke, "Huntington's Disease Information Page," http://www.ninds.nih.gov/disorders/huntington/huntington.htm.

42. A. Lasham et al., "The Y-Box-Binding Protein, YB1, Is a Potential Negative Regulator of the p53 Tumor Suppressor," *Journal of Biological Chemistry* 278, no. 37 (2003), http://www.ncbi.nlm.nih.gov/entrez/query.fcgi?cmd=Retrieve&db=PubMed&list_uids=12835324&dopt=Abstract.

43. Amir Shlomai and Yosef Shaul, "Inhibition of Hepatitis B. Virus Expression and Replication by RNA Interference," *Hepatology* (April 2003), http://www.weizmann.ac.il/molgen/members/Shaul/hep-764.pdf.

44. Erika Check, "RNA Treatment Lowers Cholesterol," *news@nature.com*,

November 10, 2004, http://www.nature.com/news/2004/041108/pf/041108-11_pf.html.

45. E. J. Mundell, "Scientists Find Key to Gene Therapy," WCNC.com, November 10, 2004, http://ww2.wcnc.com/Global/story.asp?S=2549583.

46. Check, "RNA Treatment Lowers Cholesterol."

47. GeneTests, *Gene Review*, http://www.genetests.org (accessed April 10, 2005).

48. "Huntington's Disease Genetic Testing Basics," Aetna InteliHealth, July 16, 2003, http://www.intelihealth.com/IH/ihtIH/WSIHW000/32193/35429.html.

49. Richard Willing, "Gene Tests Bring Agonizing Choices: Patients, Families, Doctors Wrestle with Privacy Issues," *USA Today*, April 23, 2001, p. 1A.

50. Stacy Singer, "How Understanding Human Genes Will Change Medicine," Cox News Service, November 8, 2004.

51. George Annas, personal communication, June 1999.

52. Gregory Stock, personal communication, June 1999.

53. Paul Oestreicher, personal communication, June 2001.

54. Diedtra Henderson, "Test Approved That Can Tailor Drug Doses," *Boston Globe*, December 31, 2004, p. D1.

55. Parentage Testing Program Unit, "Annual Report Summary for Testing in 2002," American Association of Blood Banks, November 2003, http://www.aabb.org/about_the_aabb/stds_and_accred/ptannrpt02.pdf.

56. Elsa C. Arnett, "Genetic Link: Lower Prices, Access Draw More People to DNA Tests," *Dallas Morning News*, November 1, 1998, p. 14A.

57. Trish Crawford, "B.C. Lab Cracks Genetics of Virus," *Toronto Star*, April 13, 2003, p. A01.

58. "CDC Lab Sequences Genome of New Coronavirus," news release, US Centers for Disease Control, April 13, 2003, http://www.cdc.gov/od/oc/media/pressrel/r030414.htm.

59. Leo L. M. Poon et al., "Rapid Diagnosis of a Coronavirus Associated with Severe Acute Respiratory Syndrome (SARS)," *Clinical Chemistry* 49, no. 7 (2003), http://www.clinchem.org/cgi/data/49/4/DC1/1.

60. Mark Shields, "Roche Launches Test for SARS Virus," Reuters, July 18, 2003, http://www.hivandhepatitis.com/health/071803a.html.

61. "First U.S. SARS Vaccine Trial Opens at NIH," States News Service, December 13, 2004.

62. Committee on Genomics Databases for Bioterrorism Threat Agents, *Seeking Security: Pathogens, Open Access, and Genome Databases* (Wash-

ington, DC: National Academies Press, 2004), http://www.nap.edu/execsumm_pdf/11087.pdf.

63. Ibid., p. 10.

64. Ibid.

65. Cathy Holding, "Lab on a Chip," *Scientist* (March 15, 2004), http://www.biomedcentral.com/news/20040315/02/.

66. Tabitha Powledge, "New Antibiotics—Resistance Is Futile," *PLOS Biology* 2, no. 2 (February 2004), http://www.plosbiology.org/plosonline/?request=get-document&doi=10.1371/journal.pbio.0020053.

67. "Approved Drug Blocks Deadly Anthrax Toxin," news release, University of Chicago Medical Center, February 16, 2004, http://www.eurekalert.org/pub_releases/2004-02/uocm-adb021104.php.

68. Huw Davies et al., "Profiling the Humoral Immune Response to Infection by Using Proteome Microarrays: High-Throughput Vaccine and Diagnostic Antigen Discovery," *Proceedings of the National Academy of Sciences* 102, no. 3 (2005), http://www.pnas.org/cgi/content/abstract/102/3/547.

69. "Drug Could Block SARS Infection," BBC News, February 3, 2004, http://newsvote.bbc.co.uk/mpapps/pagetools/print/news.bbc.co.uk/2/hi/health/3451809.stm.

70. David Derbyshire, "Gene Therapy Treatment Continues Despite Child Leukaemia Scare," *Telegraph* (London), October 4, 2002, p. 8.

71. Mark Henderson, "Pioneers Can't Put Safety First," *Times* (London), October 11, 2002, p. 4.

72. Ibid.

73. Steve Mitchell, "Gene Therapy Benefit Said to Outweigh Risk," United Press International, October 10, 2002.

CHAPTER 3

1. James A. Thomson et al., "Embryonic Stem Cell Lines Derived from Human Blastocysts," *Science* 282, no. 5391 (November 6, 1998), http://www.sciencemag.org/cgi/content/full/282/5391/1145.

2. John Gearhart, "New Potential for Human Embryonic Stem Cells," *Science* 282, no. 5391 (November 6, 1998), http://www.sciencemag.org/cgi/content/full/282/5391/1061.

3. Harold Varmus, "NIH Director's Statement on Research Using Stem Cells," December 2, 1998, http://www.nih.gov/policy/statements/120298.asp.

4. William Haseltine, remarks at the Third Annual Conference of the Society of Regenerative Medicine, November 17, 2003.

5. Paul Johnson, "Will Genetic Engineering Produce a Master Race and a Servile Multitude?" *Spectator*, March 6, 1999, p. 29.

6. "New Laws for New Tech: New Biotechnologies Should Not Be Controlled by Overly Broad Legislation," *Los Angeles Times*, November 9, 1998, p. B4.

7. Harriet Rabb, letter to Harold Varmus, January 15, 1999, http://guweb.georgetown.edu/research/nrcbl/documents/rabbmemo.pdf.

8. Do No Harm: The Coalition of Americans for Research Ethics, "On Human Embryos and Stem Cell Research," July 1, 1999, http://www.stemcellresearch.org/statement/statement.htm.

9. Edmund Pellegrino, "Public Testimony at the Release of the Do No Harm Founding Statement," July 1, 1999, http://www.stemcellresearch.org/testimony/Pellegrino.htm.

10. Yuehua Jiang et al., "Pluripotency of Mesenchymal Stem Cells Derived from Adult Marrow," *Nature* 418, no. 6893 (2002), http://www.nature.com/cgi-taf/DynaPage.taf?file=/nature/journal/v418/n6893/full/nature00870-fs.html.

11. Evan Snyder, personal communication, November 30, 2004.

12. Manuel Alvarez-Dolado et al., "Fusion of Bone-Marrow-Derived Cells with Purkinje Neurons, Cardiomyocytes and Hepatocytes," *Nature* 425, no. 6961 (2003), http://www.nature.com/cgi-taf/DynaPage.taf?file=/nature/journal/v425/n6961/abs/nature02069_r.html&dynoptions. See also David Perlman, "Marrow Stem Cells Have Limited Use, Study Finds Restorative Qualities Observed Only in Embryonic Matter," *San Francisco Chronicle*, October 16, 2003, http://www.sfgate.com/cgi-bin/article.cgi?file=/chronicle/archive/2003/10/16/MNGBH2CEI81.DTL.

13. Calvin B. Harley and Mahendra S. Rao, "Human Embryonic vs. Adult Stem Cells for Transplantation Therapies," in *Human Embryonic Stem Cells*, ed. A. Chiu and M. S. Rao (Totowa, NJ: Humana Press, 2003), pp. 239–64.

14. Roxanne Khamsi, "'Heart-Renewing' Cells Discovered," *news@nature.com*, February 9, 2005, http://www.nature.com/news/2005/050207/full/050207-8.html.

15. National Marrow Donor Program, "Umbilical Cord Blood Transplantation," http://www.marrow.org/PATIENT/cord_blood_transplantation.html.

16. "Umbilical Cord Blood Can Help Treat Adults with Leukemia: Study," CBC Health & Science News, November 24, 2004, http://www.cbc.ca/story/science/national/2004/11/24/leukemia041124.html.

17. Dick Foster, "Pioneering Transplant a Success: Doctors Confirm Anemic Girl Recovering after Gift of Newborn's Stem Cells," *Rocky Mountain News*, October 19, 2000, p. 4A.

18. "Genetic Testing of Embryos to Pick 'Savior Sibling' Okay with Most Americans; New Survey Explores Attitudes on PGD," *U.S. Newswire*, May 3, 2004.

19. Kim Tae-gyu, "Korean Scientists Succeed in Stem Cell Therapy," *Korea Times*, November 26, 2004, http://times.hankooki.com/lpage/200411/kt2004112617575710440.htm.

20. Snyder, personal communication, November 30, 2004.

21. Alexandra Goho, "Stem Cells Enable Paralysed Rats to Walk," *New Scientist* (July 3, 2003), http://www.newscientist.com/news/news.jsp?id=ns 99993894.

22. "Intravenous Infusions of Human Umbilical Cord Blood Stem Cells Benefit Rodents with ALS, Spinal Cord Injury," news release, University of South Florida Health Sciences Center, June 12, 2003, http://www.sciencedaily.com/releases/2003/06/030617081312.htm.

23. S. Saporta et al., "Human Umbilical Cord Blood Stem Cells Infusion in Spinal Cord Injury: Engraftment and Beneficial Influence on Behavior," *Journal of Hematotherapy and Stem Cell Research* 12, no. 3 (June 2003), http://www.ncbi.nlm.nih.gov/entrez/query.fcgi?cmd=Retrieve&db=PubMed&list_uids=12857368&dopt=Abstract.

24. "Cord Blood Cells Improve Rats' Neurological Recovery from Brain Injury, New Study Finds," news release, University of South Florida Health Sciences Center, June 6, 2002, http://www.sciencedaily.com/releases/2002/06 /020606073911.htm.

25. "Cord Blood Claims Questioned," *news@nature.com*, February 12, 2003, http://www.vetscite.org/publish/items/001049.

26. Ibid.

27. Snyder, personal communication, November 30, 2004.

28. Kathy E. Mitchell, "Umbilical Cord Matrix, a Rich New Stem Cell Source, Study Shows," news release, Kansas State University, January 16, 2003, http://www.eurekalert.org/pub_releases/2003-01/ksu-ucm011603.php. See also Kathy E. Mitchell et al., "Matrix Cells from Wharton's Jelly Form Neurons and

Glia," *Stem Cells* 21 (January 2003), http://stemcells.alphamedpress.org/cgi/content/abstract/21/1/50.

29. Commission on Life Sciences, *Stem Cells and the Future of Regenerative Medicine* (Washington, DC: National Academy of Sciences Press, 2002), p. 59. See also Rick Weiss, "Broader Stem Cell Research Backed," *Washington Post*, September 11, 2002, http://www.washingtonpost.com/ac2/wp-dyn/A7126-2001Sep11.

30. Theresa Tamkins, "Human Embryos Cloned," *Scientist* (February 14, 2004), http://www.biomedcentral.com/news/20040212/02/.

31. Philip Cohen, "Therapeutic Cloning 'Proof of Principle,'" *New Scientist* (June 2, 2002), http://www.newscientist.com/news/news.jsp?id=ns99992356.

32. Marshall Eliot, "Claim of Human-Cow Embryo Greeted with Skepticism," *Science* 283, no. 5393 (November 20, 1998): 1390.

33. "Scientists Produce Cloned Hybrid Embryo," *Korea Times*, March 9, 2002.

34. Rick Weiss, "Rabbit-Human Embryos Reported: Chinese Scientists Mix DNA to Create Source of Stem Cells," *San Francisco Chronicle*, August 14, 2003, http://www.sfgate.com/cgi-bin/article.cgi?f=/c/a/2003/08/14/MN309466.DTL.

35. Roger Pederson, personal communication, September 30, 2004.

36. Eric Cohen and William Kristol, "Should Human Cloning Be Allowed? No, It's a Moral Monstrosity," *Wall Street Journal*, December 5, 2001, http://www.newamerica.net/index.cfm?pg=article&DocID=656.

37. J. Bottum, "Animal Planet," *Weekly Standard*, December 6, 2001, http://www.weeklystandard.com/Content/Public/Articles/000/000/000647nqenv.asp.

38. Daniel Haney, "Scientists Choose a Transplant Donor That's Smart, Plentiful, and Kind of Cute," *Boston Globe*, August 4, 2001, http://www.boston.com/news/daily/04/pig_transplant.htm.

39. Bottum, "Animal Planet."

40. David Longtin and Duane Kraemer, "Concerns over Human-Animal Experiments Overblown," *USA Today*, August 10, 1999, p. 15A.

41. Sylvia Pagán Westphal, "Embryonic Stem Cells Turned into Eggs," *New Scientist* (May 3, 2003), http://www.newscientist.com/news/news.jsp?id=ns99993688.

42. Wesley J. Smith, "The False Promise of 'Therapeutic' Cloning," *Weekly Standard*, March 11, 2002, http://www.weeklystandard.com/Content/Public/Articles/000/000/000/972qujyx.asp.

43. Judy Norsigian, statement before the Senate Health, Education, Labor,

and Pensions Committee, March 5, 2002, http://www.ourbodiesourselves.org/clone4.htm.

44. Thomas Okarma, personal communication, June 2004.

45. Ibid.

46. Ibid.

47. Gideon Gil, "U of L Trial Targets Organ Rejection; Aim Is to Trick Immune System without Drugs," *Louisville Courier Journal*, April 5, 2003, p. 1B.

48. Sylvia Pagán Westphal, "'Humanised' Organs Can Be Grown in Animals," *New Scientist* (December 17, 2003), http://www.newscientist.com/hottopics/cloning/cloning.jsp?id=ns99994492.

49. Rick Weiss, "Of Mice, Men and In-Between," *Washington Post*, November 19, 2004, http://www.washingtonpost.com/ac2/wp-dyn/A63731-2004Nov19.

50. Council of Europe Working Party on Xenotransplantation, *Report on the State of the Art in the Field of Xenotransplantation*, February 21, 2003, http://www.coe.int/T/E/Legal_Affairs/Legal_co-operation/Bioethics/Activities/Xenotransplantation/XENO(2003)1_SAR.pdf.

51. Bryn Nelson, "Alive: With Help of a Pig's Liver," *Newsday*, August 19, 2002, http://www.newsday.com/news/health/ny-liverpart3.story.

52. Ceci Connolly, "Calif. Stem Cell Initiative Could Backfire Nationally," *Washington Post*, November 14, 2004, http://www.washingtonpost.com/wp-dyn/articlesA48184Nov13.html.

53. Steve Quayle, "Chimera Patent Application Holds Mirror to Biotech Society," *Genewatch*, 1999, http://www.stevequayle.com/News.alert/Genetic_Manip/020418.Chimera.patent.html.

54. Mark Dowie, "Gods and Monsters," *Mother Jones* (January/February 2004), http://www.motherjones.com/news/feature/2004/01/12_401.html.

55. Ibid.

56. Jenna Green, "He's Not Just Monkeying Around," *Legal Times*, August 16, 1999, p. 17.

57. Maggie Fox, "U.S. Fertility Group Offers Embryo Stem Cells," Reuters, January 27, 2005.

58. "Scientists Track Down the Root of Cloning Problems," news release, Whitehead Institute for Biomedical Research, May 14, 2001, http://www.sciencedaily.com/releases/2001/05/010511073756.

59. Lawrence Goldstein, remarks at Biotechnology Industry Organization Annual Conference, June 24, 2003.

60. George W. Bush, "Remarks by the President on Stem Cell Research," August 9, 2001, http://www.whitehouse.gov/news/releases/2001/08/20010809-2.html.

61. Gemma Booth, "The Stem Cell Refugee," *Wired*, December 2003, http://www.wired.com/wired/archive/11.12/start.html?pg=14.

62. "Those Favoring Stem Cell Research Increases to a 73 to 11 Percent Majority," Harris Poll, August 18, 2004, http://www.harrisinteractive.com/harris_poll/index.asp?PID=488.

63. "Momentum Grows for Stem Cell Legislation after Year of Change," *Boston Herald*, November 19, 2004, http://www.bostonherald.com/localPolitics/view.bg?articleid=54916.

64. Adam Clymer, "Public Favors Stem Cell Research, Annenberg Polling Data Show Press Release," National Annenbert Election Survey, August 9, 2004, http://www.annenbergpublicpolicycenter.org/naes/2004_03_stem-cell_08-09_pr.pdf.

65. Todd Ackerman, "Support Growing in US for Embryonic Stem-Cell Research as 53% Approve," *Financial Express*, October 14, 2004, http://www.financialexpress.com/fe_full_story.php?content_id=71349.

66. Shaoni Bhattacharya, "UK Scientists to Clone Human Embryos," *New Scientist* (August 11, 2004), http://www.newscientist.com/news/news.jsp?id=ns99996272.

67. Megan Garvey, "California Stem Cell Project Energizes Other States to Act," *Los Angeles Times*, November 22, 2004, p. A1.

68. B. Vastag, "Private Efforts Pick Up Stem Cell Slack," *Journal of the American Medical Association* 291, no. 17 (2004): 2059–60.

69. Joseph Cardinal Ratzinger, *Instruction on Respect for Human Life in Its Origin and on the Dignity of Procreation: Replies to Certain Questions of the Day*, February 22, 1987, http://www.vatican.va/roman_curia/congregations/cfaith/documents/rc_con_cfaith_doc_19870222_respect-for-human-life_en.html.

70. Ethics and Religious Liberty Commission, Southern Baptist Convention, "Statement on Human Stem Cell Research," October 26, 2004, http://www.erlc.com/partner/Article_Display_Page/0,,PTID313086%7CCHID590694%7CCIID1890076,00.html.

71. Presbyterian Church (USA), "Presbyterian Statement on Stem Cell Research," June 2001, http://www.aaas.org/spp/dser/bioethics/perspectives/pstatement.shtml; Episcopal Church, "Final Resolution: Genetics: Approve

Research on Human Stem Cells," August 2003, http://www.submitresolution.dfms.org/view_leg_detail.aspx?id=A014&type=FINAL.

72. Elliot N. Dorff, "Embryonic Stem Cell Research: The Jewish Perspective," *United Synagogue Review* (Spring 2002), http://www.uscj.org/Embryonic_Stem_Cell_5809.html.

73. Abdulaziz Sachedina, "Cellular Division," *Science & Spirit* (2002), http://www.science-spirit.org/webextras/sachedina.html.

74. Muzammil Siddiqi, "An Islamic Perspective on Stem Cell Research," PakistanLink (2001), http://www.pakistanlink.com/religion/2001/0803.html.

75. Glenn McGee and Arthur Caplan, "What's in the Dish?" *Hastings Center Report* 29, no. 4 (July–August 1999), http://www.ncbi.nlm.nih.gov/entrez/query.fcgi?cmd=Retrieve&db=PubMed&list_uids=10321341&dopt=Abstract.

76. Patrick Lee and Robert P. George, "Reason, Science & Stem Cells," *National Review*, July 20, 2001, http://www.nationalreview.com/comment/comment-george073001.shtml.

77. Patrick Lee and Robert P. George, "The Stubborn Facts of Science," *National Review*, July 30, 2001, http://www.nationalreview.com/comment/comment-george073001.shtml.

78. Michael J. Sandel, "Embryo Ethics—The Moral Logic of Stem-Cell Research," *New England Journal of Medicine* 351, no. 3 (July 15, 2004), http://content.nejm.org/cgi/content/full/351/3/207.

79. John Opitz, testimony before the President's Council on Bioethics, January 16, 2003, http://www.bioethics.gov/transcripts/jan03/jan16full.html.

80. Sandel, "Embryo Ethics."

81. President's Council on Bioethics, "Session 6: Seeking Morally Unproblematic Sources of Human Embryonic Stem Cells," transcript, December 3, 2004, http://www.bioethics.gov/transcripts/dec04/session6.html.

82. Ibid.

83. D. A. Melton, G. Q. Daley, and C. G. Jennings, "Altered Nuclear Transfer in Stem-Cell Research—A Flawed Proposal," *New England Journal of Medicine* 351, no. 27 (December 20, 2004): 2792.

84. Harold Varmus, testimony before the National Bioethics Advisory Commission, January 19, 1999, http://www.georgetown.edu/research/nrcbl/nbac/transcripts/Jan99/day_1.pdf.

85. Julian Savulescu, "Should We Clone Human Beings? Cloning as a Source of Tissue for Transplantation," *Journal of Medical Ethics* 25 (April 1999): 87–95.

86. Ibid.

87. Ronald Bailey, "Petri Dish Politics," *Reason* (December 1999), http://www.reason.com/9912/fe.rb.petri.shtml.

CHAPTER 4

1. Emma Young, "First Cloned Baby 'Born on December 26,'" *New Scientist* (December 27, 2002), http://www.newscientist.com/news/news.jsp?id=ns 99993217.

2. David Adams and Thomas Tobin, "A Cloned Baby Is Born—If You Believe It," *St. Petersburg Times*, December 28, 2002, p. 1A.

3. Dana Canedy with Kenneth Chang, "Group Says Human Clone Was Born to an American," *New York Times*, December 28, 2002, p. 16.

4. Clifford Krauss, "The Clone Ranger Basks in the Glow; Claims Bring Lots of Publicity," *International Herald Tribune*, February 25, 2003, p. 6.

5. Michael Day, "'I've Implanted First Cloned Human Embryo' Claims Doctor; US Scientist Makes Outlandish Allegations in London—But Is Met with Outrage and Skepticism," *Sunday Telegraph* (London), January 18, 2004, p. 10.

6. David Derbyshire, "Grieving Parents Paid Clone Doctor for Embryo Trials; If This Is a Valid Research Project, Why Is He Using Corpses?" *Telegraph* (London), September 1, 2004, p. 6.

7. Rick Weiss, "Animals in U.S. and Europe Now Pregnant with Clones; Methods Mimic Those That Created Dolly," *Washington Post*, June 28, 1997, p. A1.

8. Sharon Schmickle, "Cloning Controversy," *Minneapolis Star Tribune*, March 13, 1997, p. 1A.

9. Ibid.

10. Joel Achenbach, "Send in the Clones," *Washington Post*, February 25, 1997, http://www.washingtonpost.com/wp-srv/national/longterm/science/cloning /cloning4.htm.

11. George Will, "The Moral Hazards of Scientific Wonders," *Washington Post*, February 26, 1997, p. A19.

12. Daniel Callahan, "A Step Too Far," *New York Times*, February 26, 1997, p. 23.

13. Rick Weiss, "Scientists Achieve Cloning Success," *Washington Post*,

February 24, 1997, http://www.washingtonpost.com/wp-srv/national/longterm/science/cloning/hellodolly.htm.

14. Gina Kolata, "Ethics Panel Recommends a Ban on Human Cloning," *New York Times*, June 8, 1997, p. 22.

15. Gina Kolata and Andrew Pollack, "A Breakthrough on Cloning? Perhaps, or Perhaps Not Yet," *New York Times*, November 27, 2001, p. 1.

16. Sheryl Gay Stolberg, "Bush Denounces Cloning and Calls for Ban," *New York Times*, November 27, 2001, p. 12.

17. President's Council on Bioethics, *Human Cloning and Human Dignity: An Ethical Inquiry* (Washington, DC: US Government Printing Office, 2002), pp. 205–66.

18. H. Tristram Engelhardt, personal communication, March 1997.

19. Will, "The Moral Hazards of Scientific Wonders."

20. President's Council on Bioethics, *Human Cloning and Human Dignity*, pp. 102–103.

21. John Robertson, "The $1000 Genome: Ethical and Legal Issues in Whole Genome Sequencing of Individuals," *American Journal of Bioethics* 3, no. 3 (Summer 2003), http://www.utexas.edu/law/faculty/jrobertson/1000genome.pdf.

22. President's Council on Bioethics, *Human Cloning and Human Dignity*, p. 99.

23. Ibid.

24. Stanley K. Henshaw, "Unintended Pregnancy in the United States," *Family Planning Perspectives* 30, no. 1 (January/February 1998), http://www.agi-usa.org/pubs/journals/3002498.html.

25. Christopher Bond, "Bond Joins Effort to Ban Human Cloning," April 26, 2001, http://bond.senate.gov/atwork/recordtopic.cfm?id=176286.

26. Scott Simon, *Weekend Saturday*, National Public Radio, March 1, 1997.

27. Rudolph Jaenisch, "Human Cloning—The Science and Ethics of Nuclear Transplantation," *New England Journal of Medicine* 351, no. 27 (December 20, 2004): 2784; Rudolph Jaenisch and Ian Wilmut, "Don't Clone Humans," *Science* 291, no. 5513 (March 30, 2001): 2552.

28. James C. Cross, "Factors Affecting the Developmental Potential of Cloned Mammalian Embryos," *Proceedings of the National Academy of Sciences* 98, no. 11 (May 22, 2001), http://www.pnas.org/cgi/content/full/98/11/5949.

29. Jaenisch, "Human Cloning," p. 2787.

30. Ibid., p. 2789.

31. Sylvia Pagán Westphal, "'Virgin Birth' Mammal Rewrites Rules of Biology," *New Scientist* (April 21, 2004), http://www.newscientist.com/article.ns?id=dn4909.

32. Helen Pearson, "Biologists Come Close to Cloning Monkeys," *news@nature.com*, October 21, 2004, http://www.nature.com/news/2004/041018/pf/041018-12_pf.html.

33. Gretchen Vogel, "Misguided Chromosomes Foil Primate Cloning," *Science* 300, no. 5617 (April 11, 2003): 225–27.

34. Engelhardt, personal communication, March 1997.

35. Ibid.

CHAPTER 5

1. Yury Verlinsky et al., "Preimplantation Diagnosis of Early Onset Alzheimer Disease Caused by V717L Mutation," *Journal of the American Medical Association* 287, no. 8 (February 27, 2002): 1018–21.

2. Myrna E. Watanabe, "Advancing Whole Genome Amplification," *Scientist* 17, no. 1 (January 13, 2003), http://www.the-scientist.com/yr2003/jan/tools_030113.html.

3. Michael Le Page, "Is a New Era Dawning for Embryo Screening?" *New Scientist* (July 24, 2004), http://archive.newscientist.com/secure/article/article.jsp?rp=1&id=mg18324571.200.

4. Robert Edwards, personal communication, September 2004.

5. James Randerson, "Babies Born after Surgery on Eggs," *New Scientist* (October 23, 2004): 12.

6. R. J. Bartlett et al., "In Vivo Targeted Repair of a Point Mutation in the Canine Systrophin Gene by a Chimeric RNA/DNA Oligonucleotide," *Nature Biotechnology* 18, no. 6 (June 18, 2000), http://www.ncbi.nlm.nih.gov/entrez/query.fcgi?cmd=Retrieve&db=PubMed&list_uids=10835598&dopt=Abstract; Haleh V. Samiei, "Genetic Surgery for Muscular Dystrophy in Golden Retrievers," *Genome News Network*, June 2, 2000, http://www.genomenewsnetwork.org/articles/06_00/muscular_dystrophy.shtml.

7. S. Vanderbyl et al., "Transfer and Stable Transgene Expression of a Mammalian Artificial Chromosome into Bone Marrow–Derived Human Mesenchymal Stem Cells," *Stem Cells* 22, no. 3 (2004), http://www.ncbi

.nlm.nih.gov/entrez/query.fcgi?cmd=Retrieve&db=PubMed&list_uids=15153609&dopt=Abstract.

8. Quanhe Yang et al., "Improving the Prediction of Complex Diseases by Testing for Multiple Disease-Susceptibility Genes," *American Journal of Human Genetics* 72 (2003), http://www.cdc.gov/genomics/info/reports/research/susceptibilityTest.html.

9. Michael Faraday, "Creative Quotations," http://www.creativequotations.com/one/907.htm.

10. Ethics Committee of the American Society for Reproductive Medicine, "Preconception Gender Selection for Nonmedical Reasons," *Fertility and Sterility* 75, no. 5 (May 2001), http://www.asrm.org/Media/Ethics/preconception gender.pdf.

11. David Karabinus, Genetics and IVF Institute, personal communication, September 2004.

12. Julian Savulescu, "In Defense of Selection of Nondisease Genes," *American Journal of Bioethics* 1, no. 1 (Winter 2001), http://miranda.ingentaselect.com/vl=5687049/cl=61/nw=1/fm=docpdf/rpsv/catchword/mitpress/15265161/v1n1/s8/p16.

13. "The Case against Sex Selection," Human Genetics Alert, December 2002, http://www.hgalert.org/sexselection.PDF.

14. Ibid.

15. Judith Daar, "Sliding the Slope Toward Human Cloning," *American Journal of Bioethics* 1, no. 1 (Winter 2001), http://caliban.ingentaselect.com/vl-2056106/cl=51/nw=1/fm=docpdf/rpsv/catchword/mitpress/15265161/v1n1/s11/p23.

16. Francis Fukuyama, *Our Posthuman Future: Consequences of the Biotechnology Revolution* (New York: Farrar, Straus & Giroux, 2002), p. 16.

17. Bill McKibben, "Do We Want Science to Re-Design Human Aging?" SAGE Crossroads, March 27, 2003, http://www.sagecrossroads.net/public/webcasts/02/.

18. Bill McKibben, *Enough: Staying Human in an Engineered Age* (New York: Times Books, 2003), p. 14.

19. Ibid., p. 17.

20. Helen Briggs, "Your Genetic Code on a Disc," BBC News, September 23, 2002, http://news.bbc.co.uk/2/hi/science/nature/2276095.stm.

21. McKibben, *Enough*, p. 191.

22. Ibid., p. 171.

23. Ibid., p. 172.

24. Ibid., p. 40.

25. Fukuyama, *Our Posthuman Future*, p. 9.

26. Marcy Darnovsky, "The New Eugenics: The Case against Genetically Modified Humans," *Different Takes* 4 (Spring 2000), http://www.genetics-and-society.org/resources/cgs/2000_difftakes_darnovsky.html.

27. George J. Annas, Lori B. Andrews, and Rosario M. Isasi, "Protecting the Endangered Human: Toward an International Treaty Prohibiting Cloning and Inheritable Alterations," *American Journal of Law and Medicine* 28, no. 2/3 (2002), http://www.genetics-and-society.org/resources/items/2002/ajlm_annasetal.pdf.

28. Cecilia D'Alterio and Mariano Loza-Coll, "The Homologous Origin of Analogous Structures: A Tale of an Eye for an Eye in Evolutionary Biology," *Hypothesis* 1, no. 1 (November 2003), http://medbiograd.sa.utoronto.ca/pdfs/vol1num1/Thehomologousorigin.pdf.

29. Alec MacAndrew, "What Does the Mouse Genome Draft Tell Us about Evolution?" Mouse Genome Home Page, March 1, 2003, http://www.evolutionpages.com/Mouse%20genome%20home.htm.

30. Linda MacDonald Glenn, "When Pigs Fly: Legal and Ethical Issues in Transgenics and the Creation of Chimeras," *Physiologist* 46, no. 5 (October 2003), http://www.the-aps.org/publications/tphys/2003html/Oct03/randall.htm.

31. Helen Briggs, "First Language Gene Discovered," BBC News, August 14, 2002, http://news.bbc.co.uk/1/hi/sci/tech/2192969.stm.

32. "Of Mice and Morons? Human Brain Cells in Mice," Norfolk Genetic Information Network, February 1, 2001, http://ngin.tripod.com/279.stm.

33. McKibben, *Enough*, p. 190.

34. Ronald Dworkin, *Sovereign Virtue: The Theory and Practice of Equality* (Cambridge, MA: Harvard University Press, 2000), pp. 427–52; Paul Wolfson, "Politics in a Brave New World," *Public Interest* (Winter 2001), http://www.thepublicinterest.com/archives/2001winter/article2.html.

35. Robert Wright, "Who Gets the Good Genes?" *Time*, January 11, 1999, p. 67.

36. McKibben, *Enough*, pp. 211–12.

37. Marc Sagoff, "Are Human Beings Co-creators of Nature," remarks at the Association for Politics and the Life Sciences Annual Conference, September 3, 2003.

38. McKibben, *Enough*, p. 219.

39. Ibid., p. 187.

40. Dinesh D'Souza, "Staying Human: The Danger of Techno-Utopia," *National Review*, January 22, 2001, http://www.findarticles.com/p/articles/mi_m1282/is_1_53/ai_69240190.

41. Lindsey Tanner, "Woman Gives Birth after Pre-pregnancy Test Is Used to Scan for Early Alzheimer's Gene," Associated Press, February 26, 2002.

42. Rick Weiss, "Alzheimer's Gene Screened from Newborn," *Washington Post*, February 27, 2002, p. A01.

43. Tanner, "Woman Gives Birth after Pre-pregnancy Test Is Used to Scan for Early Alzheimer's Gene."

44. Ibid.

45. Savulescu, "In Defense of Selection of Nondisease Genes."

CHAPTER 6

1. Ranjit Devraj, "Cyclone Victims Are Guinea Pigs for Mutant Food," *Asia Times*, June 13, 2000, http://www.atimes.com/ind-pak/BF13Df01.html.

2. Ronald Bailey, "Dr. Strangelunch: Or Why We Should Learn to Stop Worrying and Love Genetically Modified Foods," *Reason* (January 2001), http://reason.com/010/fe.rb.dr.shtml.

3. Ibid.

4. Meron Tesfa Michael, "Africa Bites the Bullet on Genetically Modified Food Aid," *World Press Review*, September 26, 2002, http://www.worldpress.org/Africa/737.cfm.

5. Justin Gillis, "Debate Grows over Biotech Food; Efforts to Ease Famine in Africa Hurt by U.S., European Dispute," *Washington Post*, November 30, 2003, p. A01.

6. Bailey, "Dr. Strangelunch."

7. Ibid.

8. Ibid.

9. Mike Toner, "What's Coming to Dinner? Earth Liberation Front Goes after Gene Research," *Atlanta Journal-Constitution*, May 19, 2002, p. 5F.

10. Bailey, "Dr. Strangelunch."

11. Jorge Fernandez-Cornejo and William D. McBride, "Adoption and Pesticide Use," in *Adoption of Bioengineered Crops*, USDA Economic Research Service, May 2002, http://www.ers.usda.gov/publications/aer810/aer810h.pdf.

12. Pesticide Action Network, Pesticides Database, http://www.pesticideinfo.org/List_ChemicalsAlpha.jsp.

13. USDA Economic Research Service, "Adoption of Genetically Engi-

neered Crops in the U.S.," August 20, 2004, http://www.ers.usda.gov/data /BiotechCrops.

14. Clive James, *Executive Summary: Global Status of Commercialized Biotech/GM Crops: 2004*, ISAAA Briefs No. 32 (Ithaca, NY: International Service for the Acquisition of Agri-Biotech Applications, 2004), pp. 1–12.

15. "Biotechnology Helps Protect U.S. Food Crops from Pests," news release, National Center for Food and Agricultural Policy, June 10, 2002, http://www.ncfap.org/40CaseStudies/FinalNCFAPpressrelease-221%20.pdf.

16. Janet E. Carpenter, *Case Studies in the Benefits and Risks of Agricultural Biotechnology: Roundup Ready Soybeans and Bt Corn* (Washington, DC: National Center for Food and Agricultural Policy, 2001), p. 1, http://www.ncfap .org/reports/biotech/benefitsandrisks.pdf. See also American Farm Bureau Federation, "Biotechnology's Promise Can Meet 21st-Century Needs," November 19, 2001, http://www.fb.com/news/fbn/html/biotechnology .html.

17. "ASA Study Confirms Environmental Benefits of Biotech Soybeans," news release, American Soybean Association, November 12, 2001, http://biotech-info.net/ASA_biotech_soybeans.html.

18. R. H. Phipps and J. R. Park, "Environmental Benefits of Genetically Modified Crops: Global and European Perspectives on Their Ability to Reduce Pesticide Use," *Journal of Animal and Feed Sciences* 11 (2002): 1–18.

19. Janet E. Carpenter and Leonard P. Gianessi, *Agricultural Biotechnology: Updated Benefit Estimates* (Washington, DC: National Center for Food and Agricultural Policy, 2001), p. 1, http://www.ncfap.org/reports/biotech/updatedbenefits.pdf.

20. Carpenter, *Case Studies in the Benefits and Risks of Agricultural Biotechnology*; American Farm Bureau Federation, "Biotechnology's Promise Can Meet 21st-Century Needs."

21. Ronald Bailey, "Seattle Comes to Sacramento," *Reason* (June 18, 2003), http://www.reason.com/rb/rb061803.shtml.

22. "U.S. Regulatory System Needs Adjustment as Volume and Mix of Transgenic Plants Increase in Marketplace," news release, National Academy of Sciences, April 5, 2000, http://www4.nationalacademies.org/news.nsf/isbn/0309069300 ?OpenDocument.

23. National Public Radio, "Promise and Pitfalls of Using Genetically Modified Crops in Developing Countries," *Talk of the Nation/Science*, April 14, 2000, http://www.biotech-info.net/NPR_debate.html.

24. Martina McGloughlin, "Can Biotechnology Help Fight World Hunger?"

remarks at Congressional Hunger Center Seminar, June 29, 2000, http://www.consumerscouncil.org/gmo/confprog.htm#mcgloughlin.

25. US House of Representatives, Subcommittee on Basic Research, *Seeds of Opportunity*, April 13, 2000, p. 37, http://www.agbios.com/articles/smithreport_041300.pdf.

26. Ibid.

27. European Commission, "EC-Sponsored Research on the Safety of Genetically Modified Organisms: A Review of Results," Europa, 2001, http://www.europa.eu.int/comm/research/quality-of-life/gmo/index.html#text.

28. Ivan Noble, "GM Foods 'Not Harmful,'" BBC News, May 8, 2003, http://news.bbc.co.uk/2/hi/science/nature/3007573.stm.

29. Richard Flavell, "Biotechnology and Food Nutrition Needs," in *2020 Focus No. 2: Biotechnology for Developing-Country Agriculture: Problems and Opportunities* (Washington, DC: International Food Policy Research Institute, 1999), http://www.ifpri.cgiar.org/2020/focus/focus02/focus02_02.htm.

30. Food and Agriculture Organization of the United Nations (FAO), "The State of Food and Agriculture 2003–2004," flyer, June 2004, http://www.fao.org/es/esa/pdf/sofa_flyer_04_en.pdf.

31. Felicia Wu and William B. Butz, *The Future of Genetically Modified Crops: Lessons from the Green Revolution* (Santa Monica, CA: RAND Corporation, 2004), p. xxi, http://www.rand.org/pubs/monographs/2004/RAND_MG161.pdf.

32. Lynn Cook, "Millions Served," *Forbes*, December 23, 2002, http://www.forbes.com/global/2002/1223/064.html.

33. N. Gopal Raj, "Bt Cotton, a Boon to Indian Farmers," *Hindu*, March 28, 2002, http://www.hinduonnet.com/thehindu/2002/03/28/stories/2002032802411300.htm.

34. John Whitefield, "Transgenic Cotton a Winner," *Nature* (February 7, 2003), http://www.monsanto.de/biotechnologie/publikationen/030203-12.pdf.

35. "Monsanto to Share Technologies with Danforth Center to Support Global Cassava Research," news release, Donald Danforth Plant Science Center, April 16, 2002, http://www.danforthcenter.org/newsmedia/NewsDetail.asp?nid=66.

36. Luis Herrera-Estrella, "Transgenic Plants for Tropical Regions: Some Considerations about Their Development and Their Transfer to the Small Farmer" (National Academy of Sciences colloquium, University of California, Irvine, December 5–6, 1998).

37. "Stress Relief: Engineering Rice Plants with Sugar-Producing Gene Helps Them Tolerate Drought, Salt and Low Temperatures, Cornell Biologists

Report," news release, Cornell University, November 25, 2002, http://www.news.cornell.edu/releases/Nov02/trehalose_stress.hrs.html.

38. "Market Boost for Trehalose Sweetener," FOOD navigator.com/europe, January 20, 2004, http://www.foodnavigator.com/news/pringNewsBis.asp?id=49187.

39. McGloughlin, "Can Biotechnology Help Fight World Hunger?"

40. N. Seppa, "Edible Vaccine Spawns Antibodies to Virus," *Science News*, July 22, 2000, http://www.sciencenews.org/20000722/fob6.asp.

41. Pew Initiative on Food and Biotechnology, *Pharming the Field: A Look at the Benefits and Risks of Bioengineering Plants to Produce Pharmaceuticals*, February 28, 2003, http://pewagbiotech.org/events/0717/ConferenceReport.pdf.

42. Ibid.

43. "Texas A&M Scientists Clone First-Ever Bull," news release, Texas A&M University, September 2, 1999, http://agnews.tamu.edu/dailynews/stories/VETM/Sep0299a.htm.

44. *Animal Biotechnology: Science Based Concerns* (Washington, DC: National Academy of Sciences Press, 2003), http://books.nap.edu/execsumm_pdf/10418.pdf.

45. Ivan Noble, "Spider Scientists Spin Tough Yarn," BBC News, January 17, 2002, http://news.bbc.co.uk/1/hi/sci/tech/1760059.stm.

46. "ALLERCA to Begin Development of Allergen Free Cats as an Alternative to Current Allergy Treatments," news release, ALLERCA, August 22, 2004, http://www.prweb.com/releases/2004/8/prweb151517.htm.

47. Food and Agriculture Organization of the United Nations, *State of the World's Fisheries and Aquaculture* (Rome: FAO, 2002), http://www.fao.org/docrep/005/y7300e/y7300e04.htm#P5_111.

48. "Biotech Project Could Protect Grapes from Disease-Spreading Bug," *Marin Independent Journal*, August 25, 2004, http://pewagbiotech.org/newsroom/summaries/display.php3?NewsID=734.

49. California Department of Agriculture, "Pink Bollworm Sterile Release," http://www.cdfa.ca.gov/phpps/ipc/pinkbollworm/pbw_sterilerelease.htm.

50. US Department of Agriculture, Forest Service, "Gypsy Moth Nucleopolyhedrosis Virus," http://infoventures.com/e-hlth/pesticide/gypchek.html.

51. "Biological Control Successful against Invasive Plant," news release, Pennsylvania State University, August 1, 2004, http://paipm.cas.psu.edu/NewsReleases/ploosestrife.html.

52. Sierra Club, *Genetic Engineering at a Historic Crossroads*, March 2001, http://www.sierraclub.org/biotech/report.asp.

53. Mark Sagoff, "What's Wrong with Exotic Species?" *Report of the Institute for Philosophy and Public Policy* 19, no. 4 (Fall 1999), http://www.puaf.umd.edu/IPPP/fall1999/exotic_species.htm.

54. Steven H. Strauss, Stephen P. DiFazio, and Richard Meilan, "Genetically Modified Poplars in Context," *Forestry Chronicle* 77, no. 2 (March/April 2001), http://wwwdata.forestry.oregonstate.edu/tgbb/articles/Strauss_2001_The_Forestry_Chronicle.pdf.

55. Scott Merkle, "American Chestnut," in *The New Georgia Encyclopedia*, http://www.georgiaencyclopedia.org/nge/Article.jsp?id=h-944.

56. Naomi Lubick, "Designing Trees," *Scientific American* (April 1, 2002), http://www.sciam.com/article.cfm?articleID=0000BA96-AE60-1CDA-B4A8809EC588EEDF.

57. "Attacks on Genetic Research Threaten Scientists, Scientific Process," *Forestry Source*, July 2001, http://www.safnet.org/archive/biotech701.cfm.

58. Sierra Club, "Genetically Engineered Trees," http://www.sierraclub.org/biotech/trees.asp.

59. Straus, DiFazio, and Meilan, "Genetically Modified Poplars in Context."

60. Ibid.

61. National Research Council, *Biological Confinement of Genetically Engineered Organisms* (Washington, DC: National Academies Press, 2004).

62. Hembree Brandon, "GMO Crop or Non-GMO? Exorcist Gene Offers Both," *Delta Farm Press*, October 4, 2002, http://deltafarmpress.com/mag/farming_gmo_crop_nongmo/; R. J. Keenan and W. P. Stemmer, "Nontransgenic Crops from Transgenic Plants," *Nature Biotechnology* (March 2002): 215–16.

63. Vandana Shiva, *Stolen Harvest: The Hijacking of the Global Food Supply* (Cambridge, MA: South End Press, 1999), p. 83.

64. RAFI, "Terminator Technology Targets Farmers," March 30, 1998, http://www.rafi.org/text/txt_article.asp?newsid=188.

65. Stephen Leahy, "Environment: World Ban on 'Terminator' Seeds Survives a Challenge," IPS–Inter Press Service, February 11, 2005.

66. USDA Agricultural Research Service, "Cloroplast Genetic Engineering," *2003 Annual Report*, http://www.ars.usda.gov/research/projects.htm?ACCN_NO=405415&showpars=true&fy=2003.

67. FAO, "The State of World Food and Agriculture 2003–2004."

68. Stanley W. B. Ewen and Arpad Pusztai, "Effect of Diets Containing Genetically Modified Potatoes Expressing *Galanthus nivalis lectin* on Rat Small Intestine," *Lancet* (October 16, 1999): 1353–54.

69. "GM Controversy Intensifies," BBC News, October 16, 1999, http://news.bbc.co.uk/1/hi/sci/tech/474911.stm.

70. Julie A. Nordlee et al., "Identification of a Brazil-Nut Allergen in Transgenic Soybeans," *New England Journal of Medicine* 334, no. 11 (March 14, 1996), http://faculty.washington.edu/sstrand/Brazil_Nut_Allergin.pdf.

71. Centers for Disease Control, National Center for Environmental Health, *Investigation of Human Health Effects Associated with Potential Exposure to Genetically Modified Corn*, June 11, 2001, http://www.cdc.gov/nceh/ehhe/Cry9c Report/cry9creport.pdf.

72. "DNA Contamination Feared," *Washington Post*, December 3, 2001, p. A09.

73. Anita Manning, "Gene-Altered DNA May Be 'Polluting' Corn," *USA Today*, November 28, 2001, http://www.usatoday.com/news/science/biology /2001-11-28-biofood-mexico.htm.

74. "Scientists Say Mexican Biodiversity Is Safe: Concerns about Cross-Pollination Unfounded," news release, Agbio World Foundation, December 21, 2001, http://ca.prweb.com/releases/2001/12/prweb31002.htm.

75. North American Commission for Environmental Cooperation, *Maize & Biodiversity: The Effects of Transgenic Maize in Mexico*, November 2004, http://www.cec.org/files/PDF/Maize-and-Biodiversity_en.pdf.

76. John E. Losey et al., "Transgenic Pollen Harms Monarch Larvae," *Nature* (May 20, 1999), http://www.biotech-info.net/transpollen.html.

77. USDA Agricultural Research Service, "Research Q&A: Bt Corn and Monarch Butterflies," http://www.ars.usda.gov/is/br/btcorn/index.html#bt1.

78. Mark K. Sears et al., "Impact of Bt Corn Pollen on Monarch Butterfly Populations: A Risk Assessment," *Proceedings of the National Academy of Sciences* (September 14, 2001), http://www.pnas.org/cgi/content/full/211329998v1.

79. Environmental Protection Agency, *Bt Plant–Pesticides Biopesticides Registration Action Document*, October 2000, p. IIC31, http://www.epa .gov/scipoly/sap/2000/October/brad3_enviroassessment.pdf.

80. G. R. Squire et al., "On the Rationale and Interpretation of the Farm Scale Evaluations of Genetically Modified Herbicide-Tolerant Crops," *Philosophical Transactions of the Royal Society of London* 358 (October 16, 2003), http://www.pubs.royalsoc.ac.uk/phil_bio/fse_content/TB031779.pdf.

81. "GM Debate Divides Farmers," *Manchester News*, October 18, 2003, http://www.manchesteronline.co.uk/news/s/70/70443_gm_debate_divides_farm ers.html.

82. "GM Crops: Blair Who Do You Represent, the British People, or Bush and the Multi-Nationals," news release, Friends of the Earth, October 16, 2003, http://www.foe.co.uk/resource/press_releases/gm_crops_blair_who_do_you.html.

83. Squire et al, "On the Rationale and Interpretation of the Farm Scale Evaluations."

84. Ibid.

85. Martina McGloughlin, *Why Safe and Effective Food Biotechnology Is in the Public Interest*, Washington Legal Foundation, Critical Legal Issues, Working Paper Series No. 99, November 1, 2000, http://www.whybiotech.com/html/con576.html.

86. Norman Ellstrand, "Gene Flow from Transgenic Crops to Wild Relatives: What Have We Learned, What Do We Know, What Do We Need to Know?" *Scientific Methods Workshop: Ecological and Agronomic Consequences of Gene Flow from Transgenic Crops to Wild Relatives*, March 5–6, 2002, http://www.agbios.com/docroot/articles/02-280-004.pdf.

87. Jodie Holt, "Prevalence and Management of Herbicide Resistant Weeds," *Scientific Methods Workshop: Ecological and Agronomic Consequences of Gene Flow from Transgenic Crops to Wild Relatives*, March 5–6, 2002, http://www.biosci.ohio-state.edu/~asnowlab/Proceedings.pdf.

88. Paul Brown, "Trade War Fear as Public Resists GM Food," *Guardian*, May 7, 2002, http://www.guardian.co.uk/gmdebate/Story/0.2763,711074,00.html.

89. "U.S. Exercises Its WTO Rights after EU Failure to Comply with WTO Rulings," news release, United States Trade Representative Office, July 12, 1999, http://www.ustr.gov/releases/1999/07/Fact.html.

90. Soren Holm and John Harris, "Precautionary Principle Stifles Discovery," *Nature* 400, no. 398 (July 28, 1999).

91. Convention on Biological Diversity, Cartagena Biosafety Protocol, http://www.biodiv.org/biosafety/protocol.asp#.

92. Commission of the European Communities, *Communication from the Commission on the Precautionary Principle*, February 2, 2000, http://europa.eu.int/comm/dqs/healthconsumer/library/pub/pub07_en.pdf.

93. Convention on Biological Diversity, Cartagena Biosafety Protocol.

94. Reuters, October 9, 2001, http://special.northernlight.com/gmfoods/end_ban.htm.

95. Tony van der Haegan, "Food Fight," remarks at Cato Institute Conference, September 25, 2002.

96. David Byrne, personal communication, October 2002.

97. Bailey, "Dr. Strangelunch."

98. USDA Agriculture Marketing Service, National Organic Program, Final Rule, http://www.ams.usda.gov/nop/nop2000/Final%20Rule/nopfinal.pdf.

99. KPMG Consulting, *Phase I Report: Economic Impact Study: Potential Costs of Mandatory Labelling of Food Products Derived from Biotechnology in Canada*, prepared for the Steering Committee of the Economic Impacts of Mandatory Food Labelling Study, University of Guelph, Ontario, December 1, 2000, http://weeds.montana.edu/news/KPMGlabelCanada.pdf.

100. USDA Economic Research Service, "Implications of Testing and Segregating Nonbiotech Crops for Grain Grades and Standards," *ERS Agriculture Information Bulletin* 762 (March 2001), http://www.ers.usda.gov/publications/aib762/aib762f.pdf.

101. Mark Condon, "Seed Purity in the Pre and Post Biotechnology Eras," *Know Where It's Going: Bringing Food to Market in the Age of Genetically Modified Crops*, September 11, 2001, http://pewagbiotech.org/events/0911/speakers/Condon.pdf.

102. Shiva, *Stolen Harvest*, p. 4.

103. Mae-Wan Ho, "Genetically Engineered Foods: The Hazards Are Inherent in the Technology," February 7, 1997, http://users.westnet.gr/~cgian/gehazards2.htm.

104. WTO, press conference, 5th Ministerial, Cancun, Mexico, September 11, 2003.

105. Bailey, "Dr. Strangelunch."

106. Shree P. Singh, "Broadening the Genetic Base of Common Bean Cultivars," *Crop Science* 41 (2001), http://crop.scijournals.org/cgi/content/full/41/6/1659.

107. Bailey, "Dr. Strangelunch."

108. Ibid.

109. "Scientists Working on 21 GM Crops," *Financial Express* (India), January 12, 2005.

110. Bailey, "Dr. Strangelunch."

CHAPTER 7

1. Paul Root Wolpe, personal communication, December 2002.
2. Jonathan Moreno, personal communication, November 2002.

3. Francis Fukuyama, *Our Posthuman Future: Consequences of the Biotechnology Revolution* (New York: Farrar, Straus & Giroux, 2002), p. 10.

4. "Open Your Mind," *Economist*, May 25, 2002.

5. William Safire, "The But-What-If Factor," *New York Times*, May 16, 2002, p. 25.

6. Peter Kramer, *Listening to Prozac* (New York: Penguin, 1997), p. 10.

7. Danielle C. Turner et al., "Cognitive Enhancing Effects of Modafinil in Healthy Volunteers," *Psychopharmacology* (January 2003), http://www.springerlink.com/app/home/contribution.asp?wasp=gaf8jb4qug3vyhc8dyvm&referrer=parent&backto=issue,8,16;journal,45,192;linkingpublicationresults,1:100390,1.

8. Melissa Healy, "Sharper Minds," *Los Angeles Times*, December 20, 2004, p. F1.

9. Fukuyama, *Our Posthuman Future*, p. 46.

10. Simon Bevilacqua, "The Doctor Who Differs," *Sunday Tasmanian*, September 23, 2001.

11. Fukuyama, *Our Posthuman Future*, p. 49.

12. Ibid., pp. 51–52.

13. Nick Megibow, personal communication, December 2002.

14. Joseph Tsien, personal communication, November 2002. See also Y. P. Tang et al., "Genetic Enhancement of Learning and Memory in Mice," *Nature* 401, no. 6748 (September 2, 1999), http://www.ncbi.nlm.nih.gov/entrez/query.fcgi?cmd=Retrieve&db=PubMed&list_uids=10485705&dopt=Abstract.

15. Y. P. Tang, E. Shimizu, and J. Z. Tsien, "Do 'Smart' Mice Feel More Pain, or Are They Just Better Learners?" *Nature Neuroscience* 4, no. 2 (February 2001), http://www.ncbi.nlm.nih.gov/entrez/query.fcgi?cmd=Retrieve&db=pubmed&dopt=Abstract&list_uids=11319546.

16. "Helicon Therapeutics, Inc., Commences Phase I Human Testing of the Company's Lead Drug Candidate," news release, Helicon Therapeutics, December 10, 2004, http://www.helicontherapeutics.com/PDFs_&_supporting_files/HT-0712_Phase_1_press_release_December_10_2004.pdf.

17. Sheena A. Josselyn et al., "Long-Term Memory Is Facilitated by Camp Response Element-Binding Protein Overexpression in the Amygdala," *Journal of Neuroscience* 21, no. 7 (April 1, 2001), http://www.jneurosci.org/cgi/content/abstract/21/7/2404; Han-Ting Zhang et al., "Inhibition of the Phosphodiesterase 4 (PDE4) Enzyme Reverses Memory Deficits Produced by Infusion of the MEK Inhibitor U0126 into the CA1 Subregion of the Rat Hippocampus," *Neuropsy-*

chopharmacology 29 (2004), http://www.nature.com/npp/journal/v29/n8/pdf/1300440a.pdf.

18. Michael Day, "'Mind Viagra' Will Restore Memory of the Middle-Aged," *Telegraph* (London), July 3, 2004, http://www.telegraph.co.uk/news/main.jhtml?xml=/news/2004/03/07/wmind07.xml.

19. Tsien, personal communication, November 2002.

20. Martha J. Farah and Paul Root Wolpe, "Monitoring and Manipulating Brain Function: New Neuroscience Technologies and Their Ethical Implications," *Hastings Center Report* (May 1, 2004): 34–35.

21. Erik Parens, "Is Better Always Good? The Enhancement Project," *Hastings Center Report* (January 1998): S1.

22. Fukuyama, *Our Posthuman Future*, p. 49.

23. Michael Gazzaniga, personal communication, November 2002.

24. Parens, "Is Better Always Good?"

25. Moreno, personal communication, November 2002.

26. Parens, "Is Better Always Good?"

27. Gazzaniga, personal communication, November 2002.

28. Martha Farah, "Emerging Ethical Issues in Neuroscience," *Nature Neuroscience* 5, no. 11 (November 2002): 1129.

29. Fukuyama, *Our Posthuman Future*, p. 46.

30. Moreno, personal communication, November 2002.

31. Ibid.

32. Parens, "Is Better Always Good?"

33. Fukuyama, *Our Posthuman Future*, p. 53.

34. Ibid., p. 49.

35. Parens, "Is Better Always Good?"

36. Kramer, *Listening to Prozac*, p. 8.

37. Wolpe, personal communication, December 2002.

38. Gazzaniga, personal communication, November 2002.

39. Donald Kennedy, "Are There Things We'd Rather Not Know?" remarks at the Neuroethics: Mapping the Field Conference, Stanford University, May 13, 2002.

40. Patricia Smith Churchland, *Brainwise: Studies in Neurophilosophy* (Cambridge, MA: MIT Press, 2002), p. 201.

41. Ibid., p. 236.

42. Fukuyama, *Our Posthuman Future*, p. 208.

43. Michael Cromartie, "Our Posthuman Future: A Conversation with Francis

Fukuyama," *Books & Culture* (July/August 2002), http://www.christianitytoday.com/magazines/bc/2002/004/4.9.html.

44. Moreno, personal communication, November 2002.

45. Michael Gazzaniga, *The Ethical Brain* (New York: Dana Press, 2005), p. 84.

CONCLUSION

1. President's Council on Bioethics, *Beyond Therapy: Biotechnology and the Pursuit of Happiness* (Washington, DC: US Government Printing Office, 2003), p. 227.

2. Ibid., p. 276.

3. Ibid., p. 303.

4. President's Council on Bioethics, *Human Cloning and Human Dignity: An Ethical Inquiry* (Washington, DC: US Government Printing Office, 2002), p. 207.

5. Francis Fukuyama, *Our Posthuman Future: Consequences of the Biotechnology Revolution* (New York: Farrar, Straus & Giroux, 2002), p. 204.

6. Ibid., p. 215.

7. Shaoni Bhattacharya, "Banned 'Designer Baby" Born in Britain," *New Scientist*, June 19, 2003, http://www.newscientist.com/article.ns?id=dn3854.

8. Robin McKie, "Human Design: Children of the Revolution," *Observer*, October 26, 2003, supplement, p. 56.

9. Mike Waites, "Designer Baby Cures Brother," *Yorkshire Post*, October 22, 2004.

10. Tom Martin, "Scots Couple Take the Battle for IVF Baby Daughter to Commons," *Sunday Express*, June 20, 2004, p. 89.

11. Liza Mundy, "Present at the Creation: The Strange Saga of the Latter-Day Baby-Making Revolution," *Washington Post Book World*, February 8, 2004, p. T04.

12. Science and Environmental Health Network, Wingspread Conference on the Precatuionary Principle, January 26, 1998, http://www.sehn.org/wing.html.

13. George Annas, personal communication, February 1999.

14. Helene Guldberg, "Challenging the Precautionary Principle," *Country Guardian*, July 1, 2003, http://www.countryguardian.net/The%20Precautionary%20Principle.htm.

15. George Poste, "The New Risks to Scientific Progress," *Financial Times*, October 28, 1999, p. 7.

16. President's Council on Bioethics, *Human Cloning and Human Dignity*, p. 220.

17. Ibid., p. 292.

18. Ibid., pp. 207–208.

19. Nick Bostrom, "In Defense of Posthuman Dignity," *Bioethics* (2003), http://www.nickbostrom.com/ethics/dignity.html.

BIBLIOGRAPHY

Achenbach, Joel. "Send in the Clones." *Washington Post*, February 25, 1997. http://www.washingtonpost.com/wp-srv/national/longterm/science/cloning/cloning4.htm.

Ackerman, Todd. "Support Growing in US for Embryonic Stem-Cell Research as 53% Approve." *Financial Express*, October 14, 2004. http://www.financialexpress.com/fe_full_story.php?content_id=71349.

Adams, David, and Thomas Tobin. "A Cloned Baby Is Born—If You Believe It." *St. Petersburg Times*, December 28, 2002.

"Advexin Vaccine Plus Chemotherapy Provides Impressive Responses in Locally Advanced Breast Cancer." *Breast Cancer News*, December 2004. http://professional.cancerconsultants.com/oncology_breast_cancer_news.aspx?id=32797.

Alexander, Brian. *Rapture: How Biotechnology Became the New Religion*. New York: Basic Books, 2003.

"ALLERCA to Begin Development of Allergen Free Cats as an Alternative to Current Allergy Treatments." News release, ALLERCA, August 22, 2004. http://www.prweb.com/releases/2004/8/prweb151517.htm.

Alvarez-Dolado, Manuel, Ricardo Pardal, Jose M. Garcia-Verdugo, John R. Fike, Hyun O. Lee, Klaus Pfeffer, Carlos Lois, Sean J. Morrison, and Arturo Alvarez-Buylla. "Fusion of Bone-Marrow-Derived Cells with Purkinje Neurons, Cardiomyocytes and Hepatocytes." *Nature* 425, no. 6961 (2003).

http://www.nature.com/cgi-taf/DynaPage.taf?file=/nature/journal/v425/n6961/abs/nature02069_r.html&dynoptions.

American Farm Bureau Federation. "Biotechnology's Promise Can Meet 21st-Century Needs," November 19, 2001. http://www.fb.com/news/fbn/html/biotechnology.html.

Animal Biotechnology: Science Based Concerns. Washington, DC: National Academy of Sciences Press, 2003. http://books.nap.edu/execsumm_pdf/10418.pdf.

Annas, George J., Lori B. Andrews, and Rosario M. Isasi. "Protecting the Endangered Human: Toward an International Treaty Prohibiting Cloning and Inheritable Alterations." *American Journal of Law and Medicine* 28, no. 2/3 (2002). http://www.genetics-and-society.org/resources/items/2002/ajlm_annasetal.pdf.

"Approved Drug Blocks Deadly Anthrax Toxin." News release, University of Chicago Medical Center, February 16, 2004. http://www.eurekalert.org/pub_releases/2004-02/uocm-adb021104.php.

Arantes-Oliveira, Nuno, Javier Apfeld, Andrew Dillin, and Cynthia Kenyon. "Regulation of Life-Span by Germ-Line Stem Cells in *Caenorhabditis elegans*." *Science* 295, no. 5554 (January 18, 2002). http://www.grg.org/UCSFWorms.htm.

Arnett, Elsa C. "Genetic Link: Lower Prices, Access Draw More People to DNA Tests." *Dallas Morning News*, November 1, 1998.

"ASA Study Confirms Environmental Benefits of Biotech Soybeans." News release, American Soybean Association, November 12, 2001. http://biotech-info.net/ASA_biotech_soybeans.html.

Aschwanden, Christie. "Gene Cheats." *New Scientist* (January 15, 2000).

Athanasiou, Tom, and Marcy Darnovsky. "The Genome as Commons." *World-Watch* (July/August 2002). http://www.genetics-and-society.org/resources/cgs/200207_worldwatch_darnovsky.html.

"Attacks on Genetic Research Threaten Scientists, Scientific Process." *Forestry Source* (July 2001). http://www.safnet.org/archive/biotech701.cfm.

Austad, Steven, *Why We Age: What Science Is Discovering about the Body's Journey through Life*. New York: John Wiley & Sons, 1997.

Bailey, Ronald. "Dr. Strangelunch: Or Why We Should Learn to Stop Worrying and Love Genetically Modified Foods." *Reason* (January 2001). http://reason.com/010/fe.rb.dr.shtml.

———. "Forever Young: The New Scientific Search for Immortality." *Reason* (August 2002). http://reason.com/0208/fe.rb.forever.shtml.

———. "Intimations of Immortality." *Reason* (March 6, 1999). http://reason.com/opeds/rb030600.shtml.

———. "Petri Dish Politics." *Reason* (December 1999). http://www.reason.com/9912/fe.rb.petri.shtml.

———. "Seattle Comes to Sacramento." *Reason* (June 18, 2003). http://www.reason.com/rb/rb061803.shtml.

Bartlett, R. J., S. Stockinger, M. M. Denis, W. T. Bartlett, L. Inverardi, T. T. Le, N. thi Man, et al. "In Vivo Targeted Repair of a Point Mutation in the Canine Systrophin Gene by a Chimeric RNA/DNA Oligonucleotide." *Nature Biotechnology* 18, no. 6 (June 18, 2000). http://www.ncbi.nlm.nih.gov/entrez/query.fcgi?cmd=Retrieve&db=PubMed&list_uids=10835598&dopt=Abstract.

Barton Davis, Elisabeth R., Daria I. Shoturma, Antonio Musaro, Nadia Rosenthal, and H. Lee Sweeney. "Viral Medicated Expression of Insulin-like Growth Factor I Blocks the Aging-Related Loss of Skeletal Muscle Function." *Proceedings of the National Academy of Science* 95, no. 26 (1998). http://www.pnas.org/cgi/content/full/95/26/15603.

Best, Ben. "Vitrification in Cryonics." http://www.benbest.com/cryonics/vitrify.html.

Bevilacqua, Simon. "The Doctor Who Differs." *Sunday Tasmanian*, September 23, 2001.

Bhattacharya, Shaoni. "Banned 'Designer Baby" Born in Britain." *New Scientist* (June 19, 2003). http://www.newscientist.com/article.ns?id=dn3854.

———. "UK Scientists to Clone Human Embryos." *New Scientist* (August 11, 2004). http://www.newscientist.com/news/news.jsp?id=ns99996272.

"Biological Control Successful against Invasive Plant." News release, Pennsylvania State University, August 1, 2004. http://paipm.cas.psu.edu/NewsReleases/ploosestrife.html.

"Biotechnology Helps Protect U.S. Food Crops from Pests." News release, National Center for Food and Agricultural Policy, June 10, 2002. http://www.ncfap.org/40CaseStudies/FinalNCFAPpressrelease-221%20.pdf.

"Biotech Project Could Protect Grapes from Disease-Spreading Bug." *Marin Independent Journal* (August 25, 2004). http://pewagbiotech.org/newsroom/summaries/display.php3?NewsID=734.

Bond, Christopher. "Bond Joins Effort to Ban Human Cloning," April 26, 2001. http://bond.senate.gov/atwork/recordtopic.cfm?id=176286.

Booth, Gemma. "The Stem Cell Refugee." *Wired*, December 2003. http://www.wired.com/wired/archive/11.12/start.html?pg=14.

Bostrom, Nick. "In Defense of Posthuman Dignity." *Bioethics* (2003). http://www.nickbostrom.com/ethics/dignity.html.

Bottum, J. "Animal Planet." *Weekly Standard*, December 6, 2001. http://www.weeklystandard.com/Content/Public/Articles/000/000/000647nqenv.asp.

Branden, Hembree. "GMO Crop or Non-GMO? Exorcist Gene Offers Both." *Delta Farm Press* (October 4, 2002). http://deltafarmpress.com/mag/farming_gmo_crop_nongmo.

Briggs, Helen. "First Language Gene Discovered." BBC News, August 14, 2002. http://news.bbc.co.uk/1/hi/sci/tech/2192969.stm.

———. "Your Genetic Code on a Disc." BBC News, September 23, 2002. http://news.bbc.co.uk/2/hi/science/nature/2276095.stm.

Brown, Kathryn. "A Radical Proposal." *Scientific American Special Edition: The Science of Staying Young* 14 (2004).

Brown, Paul. "Trade War Fear as Public Resists GM Food." *Guardian*, May 7, 2002. http://www.guardian.co.uk/gmdebate/Story/0.2763,711074,00.html.

Bush, George W. "Remarks by the President on Stem Cell Research." August 9, 2001. http://www.whitehouse.gov/news/releases/2001/08/20010809-2.html.

California Department of Agriculture. "Pink Bollworm Sterile Release." http://www.cdfa.ca.gov/phpps/ipc/pinkbollworm/pbw_sterilerelease.htm.

Callahan, Daniel. "A Step Too Far," *New York Times*, February 26, 1997.

———. "Visions of Eternity." *First Things* 133 (May 2003). http://www.firstthings.com/ftissues/ft0305/articles/callahan.html.

Canedy, Dana, with Kenneth Chang. "Group Says Human Clone Was Born to an American." *New York Times*, December 28, 2002.

Carpenter, Janet E. *Case Studies in the Benefits and Risks of Agricultural Biotechnology: Roundup Ready Soybeans and Bt Corn*. Washington, DC: National Center for Food and Agricultural Policy, 2001. http://www.ncfap.org/reports/biotech/benefitsandrisks.pdf.

Carpenter, Janet E., and Leonard P. Gianessi. *Agricultural Biotechnology: Updated Benefit Estimates*. Washington, DC: National Center for Food and Agricultural Policy, 2001. http://www.ncfap.org/reports/biotech/updatedbenefits.pdf.

"The Case against Sex Selection." Human Genetics Alert, December 2002. http://www.hgalert.org/sexselection.PDF.

Cavazzana-Calvo, Marina, Salima Hacein-Bey, Geneviève de Saint Basile,

Fabian Gross, Eric Yvon, Patrick Nusbaum, Françoise Selz, et al. "Gene Therapy of Human Severe Combined Immunodeficiency (SCID)-X1 Disease" *Science* 288, no. 5466 (April 28, 2000).

"CDC Lab Sequences Genome of New Coronavirus." News release, US Centers for Disease Control, April 13, 2003. http://www.cdc.gov/od/oc/media/pressrel/r030414.htm.

Centers for Disease Control, National Center for Environmental Health. *Investigation of Human Health Effects Associated with Potential Exposure to Genetically Modified Corn*, June 11, 2001. http://www.cdc.gov/nceh/ehhe/Cry9cReport/cry9creport.pdf.

Check, Erika. "RNA Treatment Lowers Cholesterol." *news@nature.com*, November 10, 2004. http://www.nature.com/news/2004/041108/pf/041108-11_pf.html.

Chinen, Javier, and Jennifer M. Puck. "Perspectives of Gene Therapy for Primary Immunodeficiencies." *Current Opinion in Allergy and Clinical Immunology* 4, no. 6 (2004). http://www.medscape.com/viewarticle/494616.

Churchland, Patricia Smith. *Brainwise: Studies in Neurophilosophy*. Cambridge, MA: MIT Press, 2002.

Clymer, Adam. "Public Favors Stem Cell Research, Annenberg Polling Data Show Press Release." National Annenbert Election Survey, August 9, 2004. http://www.annenbergpublicpolicycenter.org/naes/2004_03_stem-cell_08-09_pr.pdf.

Cohen, Eric, and William Kristol. "Should Human Cloning Be Allowed? No, It's a Moral Monstrosity." *Wall Street Journal*, December 5, 2001. http://www.newamerica.net/index.cfm?pg=article&DocID=656.

Cohen, Philip. "Marathon Mice Can Run and Run." *New Scientist* (August 28, 2004).

———. "Therapeutic Cloning 'Proof of Principle.'" *New Scientist* (June 2, 2002). http://www.newscientist.com/news/news.jsp?id=ns99992356.

Commission of the European Communities. *Communication from the Commission on the Precautionary Principle*. February 2, 2000. http://europa.eu.int/comm/health_consumer/library/pub/pub07_en.pdf/.

Commission on Life Sciences. *Stem Cells and the Future of Regenerative Medicine*. Washington, DC: National Academy of Sciences Press, 2002.

Committee on Genomics Databases for Bioterrorism Threat Agents. *Seeking Security: Pathogens, Open Access, and Genome Databases*. Washington, DC: National Academies Press, 2004. http://www.nap.edu/execsumm_pdf/11087.pdf.

Condon, Mark. "Seed Purity in the Pre and Post Biotechnology Eras." *Know Where It's Going: Bringing Food to Market in the Age of Genetically Modified Crops*, September 11, 2001. http://pewagbiotech.org/events/0911/speakers/Condon.pdf.

Connelly, Sheila, Julie L. Andrews, Angela M. Gallo, Dawn B. Kayda, Jiahua Qian, Leon Hoyer, Michael J. Kadan, et al. "Sustained Phenotopic Correction of Murine Hemophilia A by In Vivo Gene Therapy." *Blood* 9, no. 91 (1998). http://www.bloodjournal.org/cgi/content/full/91/9/3273?ck=nck.

Connolly, Ceci. "Calif. Stem Cell Initiative Could Backfire Nationally." *Washington Post*, November 14, 2004. http://www.washingtonpost.com/wp-dyn/articlesA48184Nov13.html.

Convention on Biological Diversity. Cartagena Biosafety Protocol. http://www.biodiv.org/biosafety/protocol.asp#.

Cook, Lynn. "Millions Served." *Forbes*, December 23, 2002. http://www.forbes.com/global/2002/1223/064.html.

"Copernicus Receives $1M from Cystic Fibrosis Foundation Therapeutics to Advance Development of a Non-Viral Gene Therapy for Cystic Fibrosis." *Business Wire*, January 20, 2005.

"Cord Blood Cells Improve Rats' Neurological Recovery from Brain Injury, New Study Finds." News release, University of South Florida Health Sciences Center, June 6, 2002. http://www.sciencedaily.com/releases/2002/06/020606073911.htm.

"Cord Blood Claims Questioned." *news@nature.com*, February 12, 2003. http://www.vetscite.org/publish/items/001049.

Council of Europe Working Party on Xenotransplantation. *Report on the State of the Art in the Field of Xenotransplantation*, February 21, 2003. http://www.coe.int/T/E/Legal_Affairs/Legal_co-operation/Bioethics/Activities/Xenotransplantation/XENO(2003)1_SAR.pdf.

Crawford, Trish. "B.C. Lab Cracks Genetics of Virus." *Toronto Star*, April 13, 2003.

Cromartie, Michael. "Our Posthuman Future: A Conversation with Francis Fukuyama." *Books & Culture* (July/August 2002). http://www.christianitytoday.com/magazines/bc/2002/004/4.9.html.

Cromie, William J. "Wine Molecule Slows Aging Process." *Harvard Gazette*, September 18, 2003. http://www.news.harvard.edu/gazette/2003/09.18/12-antiaging.html.

Cross, James C. "Factors Affecting the Developmental Potential of Cloned

Mammalian Embryos." *Proceedings of the National Academy of Sciences* 98, no. 11 (May 22, 2001). http://www.pnas.org/cgi/content/full/98/11/5949.

Daar, Judith. "Sliding the Slope toward Human Cloning." *American Journal of Bioethics* 1, no. 1 (Winter 2001). http://caliban.ingentaselect.com/vl-2056106/cl=51/nw=1/fm=docpdf/rpsv/catchword/mitpress/15265161/v1n1/s11/p23.

D'Alterio, Cecilia, and Mariano Loza-Coll. "The Homologous Origin of Analogous Structures: A Tale of an Eye for an Eye in Evolutionary Biology." *Hypothesis* 1, no. 1 (November 2003). http://medbiograd.sa.utoronto.ca/pdfs/vol1num1/Thehomologousorigin.pdf.

Dangour, Alan D. "Hormones and Supplements: Do They Work?" *Journal of Gerontology: Biological Sciences* 59 (2004). http://biomed.gerontologyjournals.org/cgi/content/full/59/7/B659.

Darnovsky, Marcy. "The New Eugenics: The Case against Genetically Modified Humans." *Different Takes* 4 (Spring 2000). http://www.genetics-and-society.org/resources/cgs/2000_difftakes_darnovsky.html.

Davies, D. Huw, Xiaowu Liang, Jenny E. Hernandez, Arlo Randall, Siddiqua Hirst, Yunxiang Mu, Kimberly M. Romero, et al. "Profiling the Humoral Immune Response to Infection by Using Proteome Microarrays: High-Throughput Vaccine and Diagnostic Antigen Discovery." *Proceedings of the National Academy of Sciences* 102, no. 3 (2005). http://www.pnas.org/cgi/content/abstract/102/3/547.

Day, Michael. "'I've Implanted First Cloned Human Embryo' Claims Doctor; US Scientist Makes Outlandish Allegations in London—But Is Met with Outrage and Skepticism." *Sunday Telegraph* (London), January 18, 2004.

———. "'Mind Viagra' Will Restore Memory of the Middle-Aged." *Telegraph* (London), July 3, 2004. http://www.telegraph.co.uk/news/main.jhtml?xml=/news/2004/03/07/wmind07.xml.

de Grey, Aubrey D. N. J. "An Engineer's Approach to the Development of Real Anti-Aging Medicine." http://www.gen.cam.ac.uk/sens/manu16.pdf.

———. "Escape Velocity: Why the Prospect of Extreme Human Life Extension Matters Now." *PLoS Biology* 2, no. 6 (2004). http://www.plosbiology.org/plosonline/?request=get-document&doi=10.1371%2Fjournal.pbio.0020187.

Derbyshire, David. "Gene Therapy Treatment Continues Despite Child Leukaemia Scare." *Telegraph* (London), October 4, 2002.

———. "Grieving Parents Paid Clone Doctor for Embryo Trials; If This Is a Valid Research Project, Why Is He Using Corpses?" *Telegraph* (London), September 1, 2004.

Devraj, Ranjit. "Cyclone Victims Are Guinea Pigs for Mutant Food." *Asia Times*, June 13, 2000. http://www.atimes.com/ind-pak/BF13Df01.html.

"DNA Contamination Feared." *Washington Post*, December 3, 2001.

Do No Harm: The Coalition of Americans for Research Ethics. "On Human Embryos and Stem Cell Research," July 1, 1999. http://www.stemcellresearch.org/statement/statement.htm.

Dorff, Elliot N. "Embryonic Stem Cell Research: The Jewish Perspective." *United Synagogue Review* (Spring 2002). http://www.uscj.org/Embryonic_Stem_Cell_5809.html.

Dowie, Mark. "Gods and Monsters." *Mother Jones*, January/February 2004. http://www.motherjones.com/news/feature/2004/01/12_401.html.

"Drug Could Block SARS Infection." BBC News, February 3, 2004. http://newsvote.bbc.co.uk/mpapps/pagetools/print/news.bbc.co.uk/2/hi/health/3451809.stm.

D'Souza, Dinesh. "Staying Human: The Danger of Techno-Utopia." *National Review*, January 22, 2001. http://www.findarticles.com/p/articles/mi_m1282/is_1_53/ai_69240190.

Duncan, David Ewing. "The Pursuit of Longevity." *Acumen Journal of Sciences* 1, no. 2 (2003).

Dworkin, Ronald. *Sovereign Virtue: The Theory and Practice of Equality*. Cambridge, MA: Harvard University Press, 2000.

"Eat What You Like, Stay Thin, Live to 120, No Cancer or Diabetes, May Be Possible Sooner Than We Think." *Medical News Today*, June 3, 2004. http://www.medicalnewstoday.com/medicalnews.php?newsid=9023.

Eliot, Marshall. "Claim of Human-Cow Embryo Greeted with Skepticism." *Science* 282, no. 5393 (November 20, 1998).

Ellstrand, Norman. "Gene Flow from Transgenic Crops to Wild Relatives: What Have We Learned, What Do We Know, What Do We Need to Know?" *Scientific Methods Workshop: Ecological and Agronomic Consequences of Gene Flow from Transgenic Crops to Wild Relatives*, March 5–6, 2002. http://www.agbios.com/docroot/articles/02-280-004.pdf.

Elston, Laura. "Sick Child's Father Defends Saviour Sibling Move." Press Association Limited, September 8, 2004.

Environmental Protection Agency. *Bt Plant–Pesticides Biopesticides Registration Action Document*, October 2000. http://www.epa.gov/scipoly/sap/2000/October/brad3_enviroassessment.pdf.

Episcopal Church. "Final Resolution: Genetics: Approve Research on Human

Stem Cells," August 2003. http://www.submitresolution.dfms.org/view_leg_detail.aspx?id=A014&type =FINAL.

Ethics and Religious Liberty Commission, Southern Baptist Convention. "Statement on Human Stem Cell Research," October 26, 2004. http://www.erlc.com/partner/Article_Display_Page/0,,PTID313086%7CCHID590694%7CCIID1890076,00.html.

European Commission. "EC-Sponsored Research on the Safety of Genetically Modified Organisms: A Review of Results." Europa, 2001. http://www.europa.eu.int/comm/research/quality-of-life/gmo/index.html#text.

Ewen, W. B. Stanley, and Arpad Pusztai. "Effect of Diets Containing Genetically Modified Potatoes Expressing *Galanthus nivalis lectin* on Rat Small Intestine." *Lancet* (October 16, 1999): 1353–54.

Farah, Martha J., and Paul Root Wolpe. "Monitoring and Manipulating Brain Function: New Neuroscience Technologies and Their Ethical Implications." *Hastings Center Report* (May 1, 2004).

Fernandez-Cornejo, Jorge, and William D. McBride. "Adoption and Pesticide Use." In *Adoption of Bioengineered Crops*. Economic Research Service, USDA, May 2002. http://www.ers.usda.gov/publications/aer810/aer810h.pdf.

Finch, Caleb, and Sarah Crimmins. "Inflammatory Exposure and Historical Changes in Human Life-Spans." *Science* 305, no. 5691 (2004).

"First U.S. SARS Vaccine Trial Opens at NIH." States News Service, December 13, 2004.

Flavell, Richard. "Biotechnology and Food Nutrition Needs." In *2020 Focus No. 2: Biotechnology for Developing-Country Agriculture: Problems and Opportunities*. Washington, DC: International Food Policy Research Institute, 1999. http://www.ifpri.cgiar.org/2020/focus/focus02/focus02_02.htm.

Food and Agriculture Organization of the United Nations. "The State of Food and Agriculture 2003–2004." Flyer, June 2004. http://www.fao.org/es/esa/pdf/sofa_flyer_04_en.pdf.

———. *State of the World's Fisheries and Aquaculture*. Rome: FAO, 2002. http://www.fao.org/docrep/005/y7300e/y7300e04.htm#P5_111.

Foster, Dick. "Pioneering Transplant a Success: Doctors Confirm Anemic Girl Recovering after Gift of Newborn's Stem Cells." *Rocky Mountain News*, October 19, 2000.

Fowler, Gene, and Bill Crawford. *Border Radio: Quacks, Yodelers, Pitchmen, Psychics, and Other Amazing Broadcasters of the American Airwaves*. New York: Limelight Editions, 1990.

Fox, Maggie. "Gene Tweaking Turns Couch Potato Mice into Racers." *USA Today*, August 27, 2004. http://www.usatoday.com/tech/science/genetics/2004-08-23-speedy-mice_X.htm.

———. "U.S. Fertility Group Offers Embryo Stem Cells." Reuters, January 27, 2005.

Freitas, Robert A. *Nanomedicine*. Austin, TX: Landes Bioscience, 1999.

Fukuyama, Francis. "In Defense of Nature, Human and Non-Human." *New Perspectives Quarterly* (July 16, 2002). http://www.digitalnpq.org/global_services/global%20viewpoint/07-16-02.html.

———. "The End of History." *National Interest* 16 (Summer 1989): 3–18.

———. *Our Posthuman Future: Consequences of the Biotechnology Revolution*. New York: Farrar, Straus & Giroux, 2002.

———. "What Are the Possibilities and the Pitfalls in Aging Research in the Future?" SAGE Crossroads, February 12, 2003. http://www.sagecrossroads.net/021203transcript.cfm.

———. "The World's Most Dangerous Ideas: Transhumanism." *Foreign Policy* (September/October 2004). http://georgeoveremire.n1/transhumanisme.n1/Fukuyama.nl.

Garvey, Megan. "California Stem Cell Project Energizes Other States to Act." *Los Angeles Times*, November 22, 2004.

Gazzaniga, Michael. *The Ethical Brain*. New York: Dana Press, 2005.

Gearhart, John. "New Potential for Human Embryonic Stem Cells." *Science* 282, no. 5391 (November 6, 1998). http://www.sciencemag.org/cgi/content/full/282/5391/1061.

"The Genesis of Gendicine: The Story behind the First Gene Therapy." *BioPharm International*, May 2004. http://www.nbsc.com/files/papers/BP5-08-04 Ae.pdf.

"Genetic Testing of Embryos to Pick 'Savior Sibling' Okay with Most Americans; New Survey Explores Attitudes on PGD." *U.S. Newswire*, May 3, 2004.

Gil, Gideon. "U of L Trial Targets Organ Rejection; Aim Is to Trick Immune System without Drugs." *Louisville Courier Journal*, April 5, 2003.

Gillis, Justin. "Debate Grows over Biotech Food; Efforts to Ease Famine in Africa Hurt by U.S., European Dispute." *Washington Post*, November 30, 2003.

Glenn, Linda MacDonald. "When Pigs Fly: Legal and Ethical Issues in Transgenics and the Creation of Chimeras." *Physiologist* 46, no. 5 (October 2003). http://www.the-aps.org/publications/tphys/2003html/Oct03/randall.htm.

"GM Controversy Intensifies." BBC News, October 16, 1999. http://news.bbc.co.uk/1/hi/sci/tech/474911.stm.

"GM Crops: Blair Who Do You Represent, the British People, or Bush and the Multi-Nationals." News release, Friends of the Earth, October 16, 2003. http://www.foe.co.uk/resource/press_releases/gm_crops_blair_who_do_you.html.

"GM Debate Divides Farmers." *Manchester News*, October 18, 2003. http://www.manchesteronline.co.uk/news/s/70/70443_gm_debate_divides_farmers.html.

Goho, Alexandra. "Stem Cells Enable Paralysed Rats to Walk." *New Scientist* (July 3, 2003). http://www.newscientist.com/news/news.jsp?id=ns99993894.

Green, Jenna. "He's Not Just Monkeying Around." *Legal Times*, August 16, 1999.

Guarente, Leonard. "Forestalling the Great Beyond with the Help of SIR2." *Scientist* (April 26, 2004). http://www.the-scientist.com/2004/4/26/34/1.

Guldberg, Helene. "Challenging the Precautionary Principle." *Country Guardian*, July 1, 2003. http://www.countryguardian.net/The%20Precautionary%20Principle.htm.

Hagen, Tory M., Jiankang Liu, Jens Lykkesfeldt, Carol M. Wehr, Russell T. Ingersoll, Vladimir Vinarsky, James C. Bartholomew, and Bruce N. Ames. "Feeding Acetyl-L-Carnitine and Lipoic Acid to Old Rats Significantly Improves Metabolic Function While Decreasing Oxidative Stress." *Proceedings of the National Academy of Sciences* 99, no. 4 (2002).

Haney, Daniel. "Scientists Choose a Transplant Donor That's Smart, Plentiful, and Kind of Cute." *Boston Globe*, August 4, 2001. http://www.boston.com/news/daily/04/pig_transplant.htm.

Harley, Calvin B., and Mahendra S. Rao. "Human Embryonic vs. Adult Stem Cells for Transplantation Therapies." In *Human Embryonic Stem Cells*, edited by A. Chiu and M. S. Rao. Totowa, NJ: Humana Press, 2003.

Hayes, Richard. "The Science and Politics of Genetically Modified Humans." *WorldWatch* (July/August 2002). http://www.genetics-and-society.org/resources/cgs/200207_worldwatch_hayes.html.

Healy, Melissa. "Sharper Minds." *Los Angeles Times*, December 20, 2004.

"Helicon Therapeutics, Inc. Commences Phase I Human Testing of the Company's Lead Drug Candidate." News release, Helicon Therapeutics, December 10, 2004. http://www.helicontherapeutics.com/PDFs_&_supporting_files/HT-0712_Phase_1_press_release_December_10_2004.pdf.

Henderson, Diedtra. "Test Approved That Can Tailor Drug Doses." *Boston Globe*, December 31, 2004.

Henderson, Mark. "Pioneers Can't Put Safety First." *Times* (London), October 11, 2002.

Henshaw, Stanley K. "Unintended Pregnancy in the United States." *Family Planning Perspectives* 30, no. 1 (January/February 1998). http://www.agi-usa.org/pubs/journals/3002498.html.

Herrera-Estrella, Luis. "Transgenic Plants for Tropical Regions: Some Considerations about Their Development and Their Transfer to the Small Farmer." National Academy of Sciences colloquium, University of California, Irvine, December 5–6, 1998.

Ho, Mae-Wan. "Genetically Engineered Foods: The Hazards Are Inherent in the Technology," February 7, 1997. http://users.westnet.gr/~cgian/gehazards2.htm.

Holding, Cathy. "Lab on a Chip." *Scientist* (March 15, 2004). http://www.biomedcentral.com/news/20040315/02/.

Holm, Soren, and John Harris. "Precautionary Principle Stifles Discovery." *Nature* 400, no. 398 (July 28, 1999).

Holt, Jodie. "Prevalence and Management of Herbicide Resistant Weeds." *Scientific Methods Workshop: Ecological and Agronomic Consequences of Gene Flow from Transgenic Crops to Wild Relatives*, March 5–6, 2002. http://www.biosci.ohio-state.edu/~asnowlab/Proceedings.pdf.

Howitz, K. T., K. J. Bitterman, H. Y. Cohen, D. W. Lamming, S. Lavu, J. G. Wood, R. E. Zipkin, et al. "Small Molecule Activators of Sirtuins Extend *Saccharomyces cerevisiae* Lifespan." *Nature* 425, no. 6954 (September 11, 2003). http://www.ncbi.nlm.nih.gov/entrez/query.fcgi?holding=npg&cmd=Retrieve&db=PubMed&list_uids=12939617&dopt=Abstract.

"Huntington's Disease Genetic Testing Basics." Aetna InteliHealth, July 16, 2003. http://www.intelihealth.com/IH/ihtIH/WSIHW000/32193/35429.html.

Hyde, S. C., D. R. Gill, C. F. Higgins, A. E. Trezise, L. J. MacVinish, A. W. Cuthbert, R. Ratcliff, M. J. Evans, and W. H. Colledge. "Correction of the Ion Transport Defect in Cystic Fibrosis Transgenic Mice by Gene Therapy." *Nature* 362, no. 6417 (1993). http://www.ncbi.nlm.nih.gov/entrez/query.fcgi?cmd=Retrieve&db=PubMed&list_uids=7681548&dopt=Citation.

"Intravenous Infusions of Human Umbilical Cord Blood Stem Cells Benefit Rodents with ALS, Spinal Cord Injury." News release, University of South Florida Health Sciences Center, June 12, 2003. http://www.sciencedaily.com/releases/2003/06/030617081312.htm.

Jaenisch, Rudolph. "Human Cloning—The Science and Ethics of Nuclear Transplantation." *New England Journal of Medicine* 351, no. 27 (December 20, 2004).

Jaenisch, Rudolph, and Ian Wilmut. "Don't Clone Humans." *Science* 291, no. 5513 (March 30, 2001).

James, Clive. *Executive Summary: Global Status of Commercialized Biotech/GM Crops: 2004.* ISAAA Briefs No. 32. Ithaca, NY: International Service for the Acquisition of Agri-Biotech Applications, 2004.

Jiang, Yuehua, Balkrishna N. Jahagirdar, R. Lee Reinhardt, Robert E. Schwartz, C. Dirk Keene, Xilma R. Ortiz-Gonzalez, Morayma Reyes, et al. "Pluripotency of Mesenchymal Stem Cells Derived from Adult Marrow." *Nature* 418, no. 6893 (2002). http://www.nature.com/cgi-taf/DynaPage.taf?file=/nature /journal /v418/n6893/full/nature00870-fs.html.

Josselyn, Sheena A., Canjun Shi, William A. Carlezon Jr., Rachael L. Neve, Eric J. Nestler, and Michael Davis. "Long-Term Memory Is Facilitated by Camp Response Element-Binding Protein Overexpression in the Amygdala." *Journal of Neuroscience* 21, no. 7 (April 1, 2001). http://www.jneurosci .org/cgi/content/abstract/21/7/2404.

Kass, Leon. "Ageless Bodies, Happy Souls: Biotechnology and the Pursuit of Perfection." *New Atlantis* 1 (Spring 2003). http://www.thenewatlantis.com /archive/1/kass/htm.

———. "The Case for Mortality." *American Scholar* 52, no. 2 (Winter 1977–78).

———. "L'Chaim and Its Limits: Why Not Immortality?" *First Things* (May 2001). http://www.firstthings.com/ftissues/ft0105/articles/kass.html.

———. "The Moral Meaning of Genetic Technology." *Commentary* (September 1999). http://www.commentarymagazine.com/Summaries/V108I2P34-1.htm.

———. "Why We Should Ban Human Cloning Now: Preventing a Brave New World." *New Republic* (May 21, 2001). http://www.tnr.com/052101/kass 052101.html.

Keenan, R. J., and W. P. Stemmer. "Nontransgenic Crops from Transgenic Plants." *Nature Biotechnology* (March 2002): 215–16.

Keene, Kimberly M., Brian D. Foy, Irma Sanchez-Vargas, Barry J. Beaty, Carol D. Blair, and Ken E. Olson. "RNA Interference Acts as a Natural Antiviral Response to O'nyong-nyong Virus (*Alphavirus*; Togaviridae) Infection of *Anopheles gambiae*." *Proceedings of the National Academy of Science* 101, no. 49 (December 7, 2004). http://www.pnas.org/cgi/content/abstract/101/49/17240.

Kennedy, Donald. "Breakthrough of the Year." *Science* 298, no. 5602 (December 20, 2002): 2283.

Kenyon, Cynthia. "Scientists Find What Type of Genes Affect Longevity." News release, University of California, San Francisco, June 29, 2003. http://pub.ucsf.edu/newsservices/releases/2003072131.

Khamsi, Roxanne. "'Heart-Renewing' Cells Discovered." *news@nature.com.* February 9, 2005. http://www.nature.com/news/2005/050207/full/050207-8.html.

Khan, Farah. "Genetics Seen to Breed Forms of Discrimination." Inter Press Service, September 4, 2001. http://www.commondreams.org/headlines01/0904-03.htm.

Khan, Shaharyar, and Rafal Smigrodzki. "A Method of Delivery, Expression, and Replacement of DNA in Mitochondria." Unpublished manuscript.

Kolata, Gina. "Ethics Panel Recommends a Ban on Human Cloning," *New York Times*, June 8, 1997.

———. "Gains on Heart Disease Leave More Survivors, and Questions." *New York Times*, January 19, 2003.

Kolata, Gina, and Andrew Pollack. "A Breakthrough on Cloning? Perhaps, or Perhaps Not Yet." *New York Times*, November 27, 2001.

Koliopoulous, Sophia. "Gene Therapy: Medicine for Your Genes." The DNA Files, October 2001. http://www.dnafiles.org/about/pgm5/topic.html.

Korn, Alison. "Genetic Test Can Guide Young Athletes." *Calgary Herald*, January 30, 2005.

KPMG Consulting. *Phase I Report: Economic Impact Study: Potential Costs of Mandatory Labelling of Food Products Derived from Biotechnology in Canada*. Prepared for the Steering Committee of the Economic Impacts of Mandatory Food Labelling Study, University of Guelph, Ontario, December 1, 2000. http://weeds.montana.edu/news/KPMGlabelCanada.pdf.

Kramer, Peter. *Listening to Prozac*. New York: Penguin, 1997.

Krauss, Clifford. "The Clone Ranger Basks in the Glow; Claims Bring Lots of Publicity." *International Herald Tribune*, February 25, 2003.

Kristol, William, and Jeremy Rifkin. "Yes: Embryonic Clones Could Lessen the Value of Human Life in U.S." *Detroit News*, July 21, 2002. http://www.detnews.com/2002/editorial.0207/22/a15-542125.htm.

Lasalandra, Michael. "Heart Patients May Ask FDA to Continue Gene Program." *Boston Herald*, July 27, 2001.

Lasham, A., S. Moloney, T. Hale, C. Homer, Y. F. Zhang, J. G. Murison, A. W. Braithwaite, and J. Watson."The Y-Box-Binding Protein, YB1, Is a Potential Negative Regulator of the p53 Tumor Suppressor." *Journal of Biological Chemistry* 278, no. 37 (2003). http://www.ncbi.nlm.nih.gov/entrez/query.fcgi ?cmd =Retrieve&db=PubMed&list_uids=12835324&dopt=Abstract.

Leahy, Stephen. "Environment: World Ban on 'Terminator' Seeds Survives a Challenge." IPS–Inter Press Service, February 11, 2005.

Lee, Patrick, and Robert P. George. "Reason, Science & Stem Cells." *National Review*, July 20, 2001. http://www.nationalreview.com/ comment/comment-george 073001.shtml.

———. "The Stubborn Facts of Science." *National Review*, July 30, 2001. http://www.nationalreview.com/comment/comment-george073001.shtml.

Le Page, Michael. "Is a New Era Dawning for Embryo Screening?" *New Scientist* (July 24, 2004). http://archive.newscientist.com/secure/article/article .jsp?rp=1&id=mg18324571.200.

"Longevity: Of Red Wine, Diet, and a Long Life." *Science STKE* (September 16, 2004). http://stke.sciencemag.org/cgi/content/abstract/sigtrans;2003/200/tw356.

Longtin, David, and Duane Kraemer. "Concerns over Human-Animal Experiments Overblown." *USA Today*, August 10, 1999.

Losordo, Douglas W., Peter R. Vale, Robert C. Hendel, Charles E. Milliken, F. David Fortuin, Nancie Cummings, Richard A. Schatz, et al. "Phase 1/2 Placebo-Controlled, Double-Blind, Dose-Escalating Trial of Myocardial Vascular Endothelial Growth Factor 2 Gene Transfer by Catheter Delivery in Patients with Chronic Myocardial Ischemia." *Circulation* 105 (2002). http://circ.ahajournals.org/cgi/content/abstract/105/17/2012?view=abstract.

Lovett, Richard A. "Gene Doping Is Shaping Up as the Next Big Issue in Debate over Performance Enhancers." *San Diego Union-Tribune*, July 21, 2004. http://www.signonsandiego.com/news/science/20040721-9999-lz 1c21doping.html.

Lu, Wei, Luiz Claudio Arraes, Wylla Tatiana Ferreira, and Jean-Marie Andrieu. "Therapeutic Dendritic-Cell Vaccine for Chronic HIV-1 Infection." *Nature Medicine* 10 (November 28, 2004). http://www.nature.com/cgi-taf/DynaPage .taf?file=/nm/journal/v10/n12/abs/nm1147.html.

Lubick, Naomi. "Designing Trees." *Scientific American* (April 1, 2002). http://www.sciam.com/article.cfm?articleID=0000BA96-AE60-1CDA-B4 A8809EC588EEDF.

Luck, Adam. "Chinese Gene Therapy Offers Hope to Terminally Ill Cancer

Patients." *Telegraph* (London), July 4, 2004. http://www.health.telegraph.co.uk/news/main.jhtml?xml=/news/2004/07/04/wgene04.xml.

MacAndrew, Alec. "What Does the Mouse Genome Draft Tell Us about Evolution?" Mouse Genome Home Page, March 1, 2003. http://www.evolutionpages.com/Mouse%20genome%20home.htm.

Manning, Anita. "Gene-Altered DNA May Be 'Polluting' Corn." *USA Today*, November 28, 2001. http://www.usatoday.com/news/science/biology/2001-11-28-biofood-mexico.htm.

Manno, Catherine, Amy J. Chew, Sylvia Hutchison, Peter J. Larson, Roland W. Herzog, Valder R. Arruda, Shing Jen Tai, et al. "AAV-Mediated Factor IX Gene Transfer to Skeletal Muscle in Patients with Severe Hemophilia B." *Blood* 101, no. 8 (2003). http://www.bloodjournal.org/cgi/content/abstract/101/8/2963?view=abstract.

"Market Boost for Trehalose Sweetener." FOOD navigator.com/europe, January 20, 2004. http://www.foodnavigator.com/news/pringNewsBis.asp?id=49187.

Martin, Tom. "Scots Couple Take the Battle for IVF Baby Daughter to Commons." *Sunday Express*, June 20, 2004.

McGee, Glenn, and Arthur Caplan. "What's in the Dish?" *Hastings Center Report* 29, no. 4 (July–August 1999). http://www.ncbi.nlm.nih.gov/entrez/query.fcgi?cmd=Retrieve&db=PubMed&list_uids=10321341&dopt=Abstract.

McGloughlin, Martina. Remarks at Congressional Hunger Center Seminar, *Can Biotechnology Help Fight World Hunger?* June 29, 2000. http://www.consumerscouncil.org/gmo/confprog.htm#mcgloughlin.

———. *Why Safe and Effective Food Biotechnology Is in the Public Interest.* Washington Legal Foundation, Critical Legal Issues, Working Paper Series No. 99, November 1, 2000. http://www.whybiotech.com/html/con576.html.

McGourty, Christine. "Eat Less to Live Longer and Healthier." *Sunday Times* (London), July 16, 2000.

McKibben, Bill. "Do We Want Science to Re-Design Human Aging?" SAGE Crossroads, March 27, 2003. http://www.sagecrossroads.net/public/webcasts/02/.

———. *Enough: Staying Human in an Engineered Age*. New York: Times Books, 2003.

McKie, Robin. "Human Design: Children of the Revolution." *Observer*, October 26, 2003.

Melton, D. A., G. Q. Daley, and C. G. Jennings. "Altered Nuclear Transfer in Stem-Cell Research—A Flawed Proposal." *New England Journal of Medicine* 351, no. 27 (December 20, 2004).

Merkle, Ralph. "Nanotechnology and Medicine." In *Advances in Anti-Aging Medicine*. Edited by Ronald M. Klatz. Larchmont, NY: Mary Ann Liebert, 1996. http://www.alcor.org/Library/html/NanotechnologyAndMedicine.html.

Merkle, Scott. "American Chestnut." In *The New Georgia Encyclopedia*. http://www.georgiaencyclopedia.org/nge/Article.jsp?id=h-944.

Michael, Meron Tesfa. "Africa Bites the Bullet on Genetically Modified Food Aid." *World Press Review*, September 26, 2002. http://www.worldpress.org/Africa/737.cfm.

Mitchell, Kathy E. "Umbilical Cord Matrix, a Rich New Stem Cell Source, Study Shows." News release, Kansas State University, January 16, 2003. http://www.eurekalert.org/pub_releases/2003-01/ksu-ucm011603.php.

Mitchell, Kathy E., Mark L. Weiss, Brianna M. Mitchell, Phillip Martin, Duane Davis, Lois Morales, Bryan Helwig, et al. "Matrix Cells from Wharton's Jelly Form Neurons and Glia." *Stem Cells* 21 (January 2003). http://stemcells.alphamedpress.org/cgi/content/abstract/21/1/50.

Mitchell, Steve. "Gene Therapy Benefit Said to Outweigh Risk." United Press International, October 10, 2002.

"Momentum Grows for Stem Cell Legislation after Year of Change." *Boston Herald*, November 19, 2004. http://www.bostonherald.com/localPolitics/view.bg?articleid=54916.

"Monsanto to Share Technologies with Danforth Center to Support Global Cassava Research." News release, Donald Danforth Plant Science Center, April 16, 2002. http://www.danforthcenter.org/newsmedia/NewsDetail.asp?nid=66.

Mount, Jane D., Roland W. Herzog, D. Michael Tillson, Susan A. Goodman, Nancy Robinson, Mark L. McCleland, Dwight Bellinger, et al. "Sustained Phenotypic Correction of Hemophilia B Dogs with a Factor IX Null Mutation by Liver-Directed Gene Therapy." *Blood* 99, no. 8 (2002). http://www.bloodjournal.org/cgi/content/abstract/99/8/2670?view=abstract.

Mundell, E. J. "Scientists Find Key to Gene Therapy." WCNC.com, November 10, 2004. http://ww2.wcnc.com/Global/story.asp?S=2549583.

Mundy, Liza. "Present at Creation: The Strange Saga of the Latter-Day Baby-Making Revolution. *Washington Post Book World*, February 8, 2004, p. T04.

Musaro, Antonio, Karl McCullagh, Angelika Paul, Leslie Houghton, Gabriella Dobrowolny, Mario Molinaro, Elisabeth R. Barton, H. L. Sweeney, and Nadia Rosenthal. "Localized Igf-1 Transgene Expression Sustains Hypertrophy and Regeneration in Senescent Skeletal Muscle." *Nature Genetics* 27, no. 2 (2001). http://www.nature.com/doifinder/10.1038/84839.

National Institute on Aging. "Alzheimer's Disease Fact Sheet," February 2003. http://www.niapublications.org/pubs/pr01-02/adfact.html.

National Institute of Neurological Disorders and Stroke. "Huntington's Disease Information Page." http://www.ninds.nih.gov/disorders/huntington/huntington.htm.

National Marrow Donor Program. "Umbilical Cord Blood Transplantation." http://www.marrow.org/PATIENT/cord_blood_transplantation.html.

National Public Radio. "Promise and Pitfalls of Using Genetically Modified Crops in Developing Countries." *Talk of the Nation/Science*, April 14, 2000. http://www.biotech-info.net/NPR_debate.html.

National Research Council. *Biological Confinement of Genetically Engineered Organisms*. Washington, DC: National Academies Press, 2004.

Nelson, Bryn. "Alive: With Help of a Pig's Liver." *Newsday*, August 19, 2002. http://www.newsday.com/news/health/ny-liverpart3.story.

"New Laws for New Tech: New Biotechnologies Should Not Be Controlled by Overly Broad Legislation." *Los Angeles Times*, November 9, 1998.

Noble, Ivan. "GM Foods 'Not Harmful.'" BBC News, May 8, 2003. http://news.bbc.co.uk/2/hi/science/nature/3007573.stm.

———. "Spider Scientists Spin Tough Yarn." BBC News, January 17, 2002. http://news.bbc.co.uk/1/hi/sci/tech/1760059.stm.

Nordhaus, William. "The Health of Nations: The Contribution of Improved Health to Living Standards," January 25, 2002. http://www.econ.yale.edu/~nordhaus/homepage/health_nber_1.doc.

Nordlee, Julie A., Steve L. Taylor, Jeffrey A. Townsend, Laurie A. Thomas, and Robert K. Bush. "Identification of a Brazil-Nut Allergen in Transgenic Soybeans." *New England Journal of Medicine* 334, no. 11 (March 14, 1996). http://faculty.washington.edu/sstrand/Brazil_Nut_Allergin.pdf.

Norsigian, Judy. Statement before the Senate Health, Education, Labor, and Pensions Committee, March 5, 2002. http://www.ourbodiesourselves.org/clone4.htm.

North American Commission for Environmental Cooperation. *Maize & Biodiversity: The Effects of Transgenic Maize in Mexico*, November 2004. http://www.cec.org/files/PDF/Maize-and-Biodiversity_en.pdf.

Oeppen, Jim, and James W. Vaupel. "Broken Limits to Life Expectancy." *Science* (May 10, 2002): 1030.

"Of Mice and Morons? Human Brain Cells in Mice." Norfolk Genetic Information Network, February 1, 2001. http://ngin.tripod.com/279.stm.

Olshansky, S. Jay, Leonard Hayflick, and Bruce A. Carnes. "The Truth about

Human Aging." *Scientific American* (May 13, 2002). http://www.sciam.com/article.cfm?chanID=sa004&articleID=0004F171-FE1E-1CDF-B4A8809EC588EEDF.

"Open Your Mind." *Economist*, May 25, 2002.

Opitz, John. Testimony before the President's Council on Bioethics, January 16, 2003. http://www.bioethics.gov/transcripts/jan03/jan16full.html.

Parens, Erik. "Is Better Always Good? The Enhancement Project." *Hastings Center Report* (January 1998).

Parentage Testing Program Unit. "Annual Report Summary for Testing in 2002." American Association of Blood Banks, November 2003. http://www.aabb.org/about_the_aabb/stds_and_accred/ptannrpt02.pdf.

Pawliuk, R., K. A. Westerman, M. E. Fabry, E. Payen, R. Tighe, E. E. Bouhassira, S. A. Acharya, et al. "Correction of Sickle Cell Disease in Transgenic Mouse Models by Gene Therapy." *Science* 294, no. 5550 (December 14, 2001). http://www.ncbi.nlm.nih.gov/entrez/query.fcgi?cmd=Retrieve&db=PubMed&list_uids=11743206&dopt=Citation.

Pearson, Helen. "Biologists Come Close to Cloning Monkeys." *news@nature.com*, October 21, 2004. http://www.nature.com/news/2004/041018/pf/041018-12_pf.html.

Pellegrino, Edmund. "Public Testimony at the Release of the Do No Harm Founding Statement." July 1, 1999. http://www.stemcellresearch.org/testimony/Pellegrino.htm.

Perlman, David. "Marrow Stem Cells Have Limited Use, Study Finds Restorative Qualities Observed Only in Embryonic Matter." *San Francisco Chronicle*, October 16, 2003. http://www.sfgate.com/cgi-bin/article.cgi?file=/chronicle/archive/2003/10/16/MNGBH2CEI81.DTL.

Pew Initiative on Food and Biotechnology. *Pharming the Field: A Look at the Benefits and Risks of Bioengineering Plants to Produce Pharmaceuticals*, February 28, 2003. http://pewagbiotech.org/events/0717/ConferenceReport.pdf.

Phipps, R. H., and J. R. Park. "Environmental Benefits of Genetically Modified Crops: Global and European Perspectives on Their Ability to Reduce Pesticide Use." *Journal of Animal and Feed Sciences* 11 (2002).

Poon, Leo L. M., On Kei Wong, Winsie Luk, Kwok Yung Yuen, Joseph S. M. Peiris, and Yi Guan. "Rapid Diagnosis of a Coronavirus Associated with Severe Acute Respiratory Syndrome (SARS)." *Clinical Chemistry* 49, no. 7 (2003). http://www.clinchem.org/cgi/data/49/4/DC1/1.

Poste, George. "The New Risks to Scientific Progress." *Financial Times*, October 28, 1999.

Powledge, Tabitha. "New Antibiotics—Resistance Is Futile." *PLOS Biology* 2, no. 2 (February 2004). http://www.plosbiology.org/plosonline/?request=get-document&doi=10.1371/journal.pbio.0020053.

Presbyterian Church (USA). "Presbyterian Statement on Stem Cell Research," June 2001. http://www.aaas.org/spp/dser/bioethics/perspectives/pstatement.shtml.

President's Council on Bioethics, *Beyond Therapy: Biotechnology and the Pursuit of Happiness*. Washington, DC: US Government Printing Office, 2003.

———. *Human Cloning and Human Dignity: An Ethical Inquiry*. Washington, DC: US Government Printing Office, 2002.

———. "Session 2: Beyond Therapy: Ageless Bodies." Transcript, March 6, 2003. http://www.bioethics.gov/transcripts/march03/session2.html.

———. "Session 6: Seeking Morally Unproblematic Sources of Human Embryonic Stem Cells." Transcript, December 3, 2004. http://www.bioethics.gov/transcripts/dec04/session6.html.

Quayle, Steve. "Chimera Patent Application Holds Mirror to Biotech Society." *Genewatch*, 1999. http://www.stevequayle.com/News.alert/Genetic_Manip/020418.Chimera.patent.html.

Rabb, Harriet. Letter to Harold Varmus, January 15, 1999. http://guweb.georgetown.edu/research/nrcbl/documents/rabbmemo.pdf.

RAFI. "Terminator Technology Targets Farmers," March 30, 1998. http://www.rafi.org/text/txt_article.asp?newsid=188.

Raj, N. Gopal. "Bt Cotton, a Boon to Indian Farmers." *Hindu*, March 28, 2002. http://www.hinduonnet.com/thehindu/2002/03/28/stories/2002032802411300.htm.

Randerson, James. "Babies Born after Surgery on Eggs." *New Scientist* (October 23, 2004).

Ratzinger, Joseph Cardinal. *Instruction on Respect for Human Life in Its Origin and on the Dignity of Procreation: Replies to Certain Questions of the Day*, February 22, 1987. http://www.vatican.va/roman_curia/congregations/cfaith/documents/rc_con_cfaith_doc_19870222_respect-for-human-life_en.html.

Rayasam, Renuka. "Introgen's Drug Enters FDA Phase: Austin Company's Gene Cancer Therapy Could Be 1st Approved in US." *Austin American-Statesman*, December 24, 2004.

Rifkin, Jeremy. "Fusion Biopolitics." *Nation*, February 18, 2002. http://www.thenation.com/doc.mhtml?i=20020218&s=rifkin.

Robertson, John. "The $1000 Genome: Ethical and Legal Issues in Whole Genome Sequencing of Individuals." *American Journal of Bioethics* 3, no. 3 (Summer 2003).

Rogers, Luis, and Steve Farrar. "'Immortal' Genes Found by Science." *Times* (London), July 4, 1999. http://www.grg.org/rose.htm.

Sachedina, Abdulaziz. "Cellular Division." *Science & Spirit* (2002). http://www.science-spirit.org/webextras/sachedina.html.

Safire, William. "The But-What-If Factor." *New York Times*, May 16, 2002.

Sagoff, Mark. "What's Wrong with Exotic Species?" *Report of the Institute for Philosophy and Public Policy* 19, no. 4 (Fall 1999). http://www.puaf.umd.edu/IPPP/fall1999/exotic_species.htm.

Salleh, Anna. "Gene Therapy Restores Sight to Blind Dogs." ABC Science Online, April 30, 2001. http://www.abc.net.au/science/news/stories/s283715.htm.

Samiei, Haleh V. "Genetic Surgery for Muscular Dystrophy in Golden Retrievers." *Genome News Network*, June 2, 2000. http://www.genomenewsnetwork.org/articles/06_00/muscular_dystrophy.shtml.

Sandel, Michael J. "Embryo Ethics—The Moral Logic of Stem-Cell Research." *New England Journal of Medicine* 351, no. 3 (July 15, 2004). http://content.nejm.org/cgi/content/full/351/3/207.

Saporta, S., J. J. Kim, A. E. Willing, E. S. Fu, C. D. Davis, and P. R. Sanberg. "Human Umbilical Cord Blood Stem Cells Infusion in Spinal Cord Injury: Engraftment and Beneficial Influence on Behavior." *Journal of Hematotherapy and Stem Cell Research* 12, no. 3 (June 2003). http://www.ncbi.nlm.nih.gov/entrez/query.fcgi?cmd=Retrieve&db=PubMed&list_uids=12857368&dopt=Abstract.

Savulescu, Julian. "In Defense of Selection of Nondisease Genes." *American Journal of Bioethics* 1, no. 1 (Winter 2001). http://miranda.ingentaselect.com/vl=5687049/cl=61/nw=1/fm=docpdf/rpsv/catchword/mitpress/15265161/v1n1/s8/p16.

———. "Should We Clone Human Beings? Cloning as a Source of Tissue for Transplantation." *Journal of Medical Ethics* 25 (April 1999).

Science and Environmental Health Network. Wingspread Conference on the Precautionary Principle, January 26, 1998. http://www.sehn.org/wing.html.

Schier, Mary Ann. "Gene Therapy Patients Celebrate Five-Year Anniversary." *Conquest* 19, no. 1 (2004): 16. http://www.mdanderson.org/pdf/conquest_2004_summer.pdf.

Schmickle, Sharon. "Cloning Controversy." *Minneapolis Star Tribune*, March 13, 1997.

"Scientists Produce Cloned Hybrid Embryo." *Korea Times*, March 9, 2002.

"Scientists Say Mexican Biodiversity Is Safe: Concerns about Cross-Pollination Unfounded." News release, Agbio World Foundation, December 21, 2001. http://ca.prweb.com/releases/2001/12/prweb31002.htm.

"Scientists Track Down the Root of Cloning Problems." News release, Whitehead Institute for Biomedical Research, May 14, 2001. http://www.sciencedaily.com/releases/2001/05/010511073756.

"Scientists Working on 21 GM Crops." *Financial Express* (India), January 12, 2005.

Sears, Mark K., Richard L. Hellmich, Diane E. Stanley-Horn, Karen S. Oberhauser, John M. Pleasants, Heather R. Mattila, Blair D. Siegfried, and Galen P. Dively. "Impact of Bt Corn Pollen on Monarch Butterfly Populations: A Risk Assessment." *Proceedings of the National Academy of Sciences* (September 14, 2001). http://www.pnas.org/cgi/content/full/211329998v1.

Secko, David. "'Longevity' Gene, Diet Linked." *Scientist* (June 18, 2004). http://www.biomedcentral.com/news/20040618/01.

Seppa, N. "Edible Vaccine Spawns Antibodies to Virus." *Science News*, July 22, 2000. http://www.sciencenews.org/20000722/fob6.asp.

Shields, Mark. "Roche Launches Test for SARS Virus." Reuters, July 18, 2003. http://www.hivandhepatitis.com/health/071803a.html.

Shiva, Vandana. *Stolen Harvest: The Hijacking of the Global Food Supply*. Cambridge, MA: South End Press, 1999.

Shlomai, Amir, and Yosef Shaul. "Inhibition of Hepatitis B. Virus Expression and Replication by RNA Interference." *Hepatology* (April 2003). http://www.weizmann.ac.il/molgen/members/Shaul/hep-764.pdf.

Siddiqi, Muzammil. "An Islamic Perspective on Stem Cell Research." PakistanLink, 2001. http://www.pakistanlink.com/religion/2001/0803.html.

Sierra Club. "Genetically Engineered Trees." http://www.sierraclub.org/biotech/trees.asp.

———. *Genetic Engineering at a Historic Crossroads*, March 2001. http://www.sierraclub.org/biotech/report.asp.

Singer, Stacy. "How Understanding Human Genes Will Change Medicine." Cox News Service, November 8, 2004.

Singh, Shree P. "Broadening the Genetic Base of Common Bean Cultivars." *Crop Science* 41 (2001). http://crop.scijournals.org/cgi/content/full/41/6/1659.

Smith, Wesley J. "The False Promise of 'Therapeutic' Cloning." *Weekly Standard*,

March 11, 2002. http://www.weeklystandard.com/Content/Public/Articles/000/000/000/972qujyx.asp.

Squire, G. R., D. R. Brooks, D. A. Bohan, G. T. Champion, R. E. Daniels, A. J. Haughton, C. Hawes, et al. "On the Rationale and Interpretation of the Farm Scale Evaluations of Genetically Modified Herbicide-Tolerant Crops." *Philosophical Transactions of the Royal Society of London* 358 (October 16, 2003). http://www.pubs.royalsoc.ac.uk/phil_bio/fse_content/TB031779.pdf.

Stein, Rob. "Muscle-Bound Boy Offers Hope for Humans: Scientists Work to Isolate Secrets of a Genetic Mutation That Could Alleviate Weakness Accompanying Disease and Aging." *Washington Post*, June 28, 2004.

Stock, Gregory, and Daniel Callahan. "Debates: Point-Counterpoint: Would Doubling the Human Life Span Be a Net Positive or a Negative for Us Either as Individuals or as a Society?" *Journal of Gerontology: Biological Sciences* 59 (2004). http://biomed.gerontologyjournals.org/cgi/content/full/59/6/B544.

Stolberg, Sheryl Gay. "The Biotech Death of Jesse Gelsinger." *New York Times Magazine*, November 30, 1999. http://www.gene.ch/gentech/1999/Dec/msg00005.html.

———. "Bush Denounces Cloning and Calls for Ban." *New York Times*, November 27, 2001.

Strauss, Steven H., Stephen P. DiFazio, and Richard Meilan. "Genetically Modified Poplars in Context." *Forestry Chronicle* 77, no. 2 (March/April 2001). http://wwwdata.forestry.oregonstate.edu/tgbb/articles/Strauss_2001_The_Forestry_Chronicle.pdf.

"Stress Relief: Engineering Rice Plants with Sugar-Producing Gene Helps Them Tolerate Drought, Salt and Low Temperatures, Cornell Biologists Report." News release, Cornell University, November 25, 2002. http://www.news.cornell.edu/releases/Nov02/trehalose_stress.hrs.html.

Swisher, Stephen J., Jack A. Roth, Ritsuko Komaki, Jian Gu, J. Jack Lee, Marshall Hicks, Jae Y., et al. "Induction of p53-Regulated Genes and Tumor Regression in Lung Cancer Patients after Intratumoral Delivery of Adenoviral p53 (INGN 201) and Radiation Therapy." *Clinical Cancer Research* 9 (January 2003). http://clincancerres.aacrjournals.org/cgi/content/abstract/9/1/93.

Tae-gyu, Kim. "Korean Scientists Succeed in Stem Cell Therapy." *Korea Times*, November 26, 2004. http://times.hankooki.com/lpage/200411/kt2004112617575710440.htm.

Tamkins, Theresa. "Human Embryos Cloned." *Scientist* (February 14, 2004). http://www.biomedcentral.com/news/20040212/02/.

Tang, Y. P., E. Shimizu, G. R. Dube, C. Rampoon, G. A. Kerchner, M. Zhuo, G. Liu, and J. Z. Tsien. "Genetic Enhancement of Learning and Memory in Mice." *Nature* 401, no. 6748 (September 2, 1999). http://www.ncbi.nlm.nih.gov/entrez/query.fcgi?cmd=Retrieve&db=PubMed&list_uids=10485705&dopt=Abstract.

Tang, Y. P., E. Shimizu, and J. Z. Tsien. "Do 'Smart' Mice Feel More Pain, or Are They Just Better Learners?" *Nature Neuroscience* 4, no. 2 (February 2001). http://www.ncbi.nlm.nih.gov/entrez/query.fcgi?cmd=Retrieve&db=pubmed&dopt=Abstract&list_uids=11319546.

Tanner, Lindsey. "Woman Gives Birth after Pre-pregnancy Test Is Used to Screen for Early Alzheimer's Gene." Associated Press, February 26, 2002.

"Texas A&M Scientists Clone First-Ever Bull." News release, Texas A&M University, September 2, 1999. http://agnews.tamu.edu/dailynews/stories/VETM/Sep0299a.htm.

Thomson, James A., Joseph Itskovitz-Eldor, Sander S. Shapiro, Michelle A. Waknitz, Jennifer J. Swiergiel, Vivienne S. Marshall, and Jeffrey M. Jones. "Embryonic Stem Cell Lines Derived from Human Blastocysts." *Science* 282, no. 5391 (November 6, 1998). http://www.sciencemag.org/cgi/content/full/282/5391/1145.

"Those Favoring Stem Cell Research Increases to a 73 to 11 Percent Majority." Harris Poll, August 18, 2004. http://www.harrisinteractive.com/harris_poll/index.asp?PID=488.

Toner, Mike. "What's Coming to Dinner? Earth Liberation Front Goes after Gene Research." *Atlanta Journal-Constitution*, May 19, 2002.

Turner, Danielle C., et al. "Cognitive Enhancing Effects of Modafinil in Healthy Volunteers." *Psychopharmacology* (January 2003). http://www.springerlink.com/app/home/contribution.asp?wasp=gaf8jb4qug3vyhc8dyvm&referrer=parent&backto=issue,8,16;journal,45,192;linkingpublicationresults,1:100390,1.

Tuschl, Thomas. "10 Emerging Technologies That Will Change Your World." *Technology Review* (February 2004). http://www.technologyreview.com/articles/04/02/emerging0204.asp?p=8.

"Umbilical Cord Blood Can Help Treat Adults with Leukemia: Study." CBC Health & Science News, November 24, 2004. http://www.cbc.ca/story/science/national/2004/11/24/leukemia041124.html.

UNESCO. Universal Declaration on the Human Genome and Human Rights.

November 11, 1997. http://portal.unesco.org/en/ev.php-URL_ID=13177&URL_DO=DO_TOPIC&URL_SECTION=201.html.

USDA Agriculture Marketing Service. National Organic Program, Final Rule. http://www.ams.usda.gov/nop/nop2000/Final%20Rule/nopfinal.pdf.

USDA Agricultural Research Service. "Cloroplast Genetic Engineering." *2003 Annual Report*. http://www.ars.usda.gov/research/projects.htm?ACCN_NO=405415&showpars=true&fy=2003.

———. "Research Q&A: Bt Corn and Monarch Butterflies." http://www.ars.usda.gov/is/br/btcorn/index.html#bt1.

USDA Economic Research Service. "Adoption of Genetically Engineered Crops in the U.S." August 20, 2004. http://www.ers.usda.gov/data/BiotechCrops.

———. "Implications of Testing and Segregating Nonbiotech Crops for Grain Grades and Standards." *ERS Agriculture Information Bulletin* 762 (March 2001). http://www.ers.usda.gov/publications/aib762/aib762f.pdf.

US Department of Agriculture, Forest Service. "Gypsy Moth Nucleopolyhedrosis Virus." http://infoventures.com/e-hlth/pesticide/gypchek.html.

"U.S. Exercises Its WTO Rights after EU Failure to Comply with WTO Rulings." News release, United States Trade Representative Office, July 12, 1999. http://www.ustr.gov/releases/1999/07/Fact.html.

US House of Representatives, Subcommittee on Basic Research. *Seeds of Opportunity*. April 13, 2000. http://www.agbios.com/articles/smithreport_041300.pdf.

"U.S. Regulatory System Needs Adjustment as Volume and Mix of Transgenic Plants Increase in Marketplace." News release, National Academy of Sciences, April 5, 2000. http://www4.nationalacademies.org/news.nsf/isbn/0309069300?OpenDocument.

Vanderbyl, S., G. N. MacDonald, S. Sidhu, L. Gung, A. Telenius, C. Perez, and E. Perkins. "Transfer and Stable Transgene Expression of a Mammalian Artificial Chromosome into Bone Marrow–Derived Human Mesenchymal Stem Cells." *Stem Cells* 22, no. 3 (2004). http://www.ncbi.nlm.nih.gov/entrez/query.fcgi?cmd=Retrieve&db=PubMed&list_uids=15153609&dopt=Abstract.

Varmus, Harold. "NIH Director's Statement on Research Using Stem Cells." December 2, 1998. http://www.nih.gov/policy/statements/120298.asp.

———. Testimony before the National Bioethics Advisory Commission. January 19, 1999. http:www.georgetown.edu/research/nrcbl/nbac/transcripts/Jan99/day_1.pdf.

Vastag, B. "Private Efforts Pick Up Stem Cell Slack." *Journal of the American Medical Association* 291, no. 17 (2004).

Verlinsky, Yury, Svetlana Rechitsky, Oleg Verlinsky, Christina Masciangelo, Kevin Lederer, and Anver Kuliev. "Preimplantation Diagnosis of Early Onset Alzheimer Disease Caused by V717L Mutation." *Journal of the American Medical Association* 287, no. 8 (February 27, 2002).

Vogel, Gretchen. "Misguided Chromosomes Foil Primate Cloning." *Science* 300, no. 5617 (April 11, 2003): 225–27.

Wadhams, Nick. "Divided U.N. Seeks Human Cloning Ban." *Washington Post*, March 8, 2005. http://washingtonpost.com/wp-dyn/articles/A17300-2005mar8.html.

Walford, Roy L, Dennis Mock, Taber MacCallum, and John L. Laseter. "Physiologic Changes in Humans Subject to Severe, Selective Calorie Restriction for Two Years in Biosphere 2: Health, Aging, and Toxicological Perspectives." *Toxicological Sciences* 52, supp. (1999). http://www.cron-web.org/Walford_Biosphere1.pdf.

Waites, Mike. "Designer Baby Cures Brother." *Yorkshire Post*, October 22, 2004.

Wang, Bing, Juan Li, and Xiao Xiao. "Adeno-Associated Virus Vector Carrying Human Minidystrophin Genes Effectively Ameliorates Muscular Dystrophy in *mdx* Mouse Model." *Proceedings of the National Academy of Sciences* 97, no. 25 (December 5, 2000). http://www.pnas.org/cgi/content/abstract/97/25/13714?view=abstract.

Wang, Yong-Xu, Chun-Li Zhang, Ruth T. Yu, Helen K. Cho, Michael C. Nelson, Corinne R. Bayuga-Ocampo, Jungyeob Ham, Heonjoong Kang, and Ronald M. Evans. "Regulation of Muscle Fiber Type and Running Indurance by PPARd." *PLoS Biology* 2, no. 10 (2004). http://www.plosbiology.org/plosonline/?request=get-document&doi=10.1371%2Fjournal.pbio.0020294.

Wanjek, Christopher. "Time in a Bottle: Anti-Aging Boosters Claim Their Products Can Turn Back the Clock. Independent Scientists Aren't Buying It." *Washington Post*, January 29, 2002.

Wasi, Safia. "RNA Interference: The Next Genetics Revolution?" *Horizon Symposia* (May 2003). http://www.nature.com/horizon/rna/background/pdf/interference.pdf.

Watanabe, Myrna E. "Advancing Whole Genome Amplification." *Scientist* 17, no. 1 (January 13, 2003). http://www.the-scientist.com/yr2003/jan/tools_030113.html.

Weale, Robert A. "Biorepair Mechanisms and Longevity." *Journal of Geron-*

tology: Biological Sciences 59 (2004). http://biomed.gerontologyjournals.org/cgi/reprint/59/5/B449.

Weiss, Rick. "Alzheimer's Gene Screened from Newborn." *Washington Post*, February 27, 2002.

———."Animals in U.S. and Europe Now Pregnant with Clones; Methods Mimic Those That Created Dolly." *Washington Post*, June 28, 1997.

———. "Boy's Cancer Prompts FDA to Halt Gene Therapy." *Washington Post*, March 4, 2005.

———."Broader Stem Cell Research Backed." *Washington Post*, September 11, 2002. http://www.washingtonpost.com/ac2/wp-dyn/A7126-2001Sep11.

———. "FDA Seeks to Penalize Gene Scientist." *Washington Post*, December 12, 2000.

———. "Of Mice, Men and In-Between." *Washington Post*, November 19, 2004. http://www.washingtonpost.com/ac2/wp-dyn/A63731-2004Nov19.

———. "Rabbit-Human Embryos Reported: Chinese Scientists Mix DNA to Create Source of Stem Cells." *San Francisco Chronicle*, August 14, 2003. http://www.sfgate.com/cgi-bin/article.cgi?f=/c/a/2003/08/14/MN309466.DTL.

———. "Scientists Achieve Cloning Success." *Washington Post*, February 24, 1997. http://www.washingtonpost.com/wp-srv/national/longterm/science/cloning/hellodolly.htm.

Westphal, Sylvia Pagán. "Cancer Gene Therapy Is First to Be Approved." *New Scientist* (November 28, 2003). http://www.newscientist.com/article.ns?id=dn4420.

———. "Embryonic Stem Cells Turned into Eggs." *New Scientist* (May 3, 2003). http://www.newscientist.com/news/news.jsp?id=ns99993688.

———. "'Humanised' Organs Can Be Grown in Animals." *New Scientist* (December 17, 2003). http://www.newscientist.com/hottopics/cloning/cloning.jsp?id=ns9999449.

———. "Just Add a Chromosome. . . ." *New Scientist* (June 19, 2004).

———. "'Virgin Birth' Mammal Rewrites Rules of Biology." *New Scientist*, April 21, 2004. http://www.newscientist.com/article.ns?id=dn4909.

Whitefield, John. "Transgenic Cotton a Winner." *Nature* (February 7, 2003). http://www.monsanto.de/biotechnologie/publikationen/030203-12.pdf.

"Who Says Chivalry Is Dead? Sir2 Fights against Aging in Mammals." Alzheimer Research Forum, June 17, 2004. http://www.alzforum.org/new/detail.asp?id=1031.

Will, George. "The Moral Hazards of Scientific Wonders." *Washington Post*, February 26, 1997.

Willing, Richard. "Gene Tests Bring Agonizing Choices: Patients, Families, Doctors Wrestle with Privacy Issues." *USA Today*, April 23, 2001.

Wolfson, Adam. "Politics in a Brave New World." *Public Interest* (Winter 2001). http://www.thepublicinterest.com/archives/2001winter/article2.html.

Wright, Robert. "Who Gets the Good Genes?" *Time*, January 11, 1999.

Wu, Felicia, and William B. Butz. *The Future of Genetically Modified Crops: Lessons from the Green Revolution*. Santa Monica, CA: RAND Corporation, 2004. http://www.rand.org/pubs/monographs/2004/RAND_MG161.pdf.

Yang, Quanhe, Muin J. Khoury, Lorenzo Botto, J. M. Friedman, and W. Dana Flanders. "Improving the Prediction of Complex Diseases by Testing for Multiple Disease-Susceptibility Genes." *American Journal of Human Genetics* 72 (2003). http://www.cdc.gov/genomics/info/reports/research/susceptibility Test.html.

Young, Emma. "First Cloned Baby 'Born on December 26.'" *New Scientist* (December 27, 2002). http://www.newscientist.com/news/news.jsp?id=ns 99993217.

Zarembo, Alan. "DNA May Soon Be in Play." *Los Angeles Times*, August 27, 2004.

Zhang, Han-Ting, Yu Zhao, Ying Huang, Nandakumar R. Dorairaj, L. Judson Chandler, and James M. O'Donnell. "Inhabition of the Phosphodiesterase 4 (PDE4) Enzyme Reverses Memory Deficits Produced by Infusion of the MEK Inhibitor U0126 into the CA1 Subregion of the Rat Hippocampus." *Neuropsychopharmacology* 29 (2004). http://www.nature.com/npp/journal /v29/n8/pdf/1300440a.pdf.

INDEX